The History of
LIMOUSIN
In North America

Edited by
Dale F. Runnion
June A. Runnion

In The United States

Part One	**The Introduction** by Richard Goff	**1967-1972**
Part Two	**The Development** by Dale F. Runnion	**1972-1978**
Part Three	**The Acceptance** by Dr. Greg Martin	**1978-1987**

In Canada

by Dale F. Runnion   **1968-1987**

Published by Dale F. Runnion
15641 El Lago Blvd.
Fountain Hills, Arizona 85268

in cooperation with
Louis de Neuville, Limoges, France

Written by: Richard Goff
             Dale F. Runnion
             Dr. Greg Martin

ISBN 0 - 9617842 - 1 - 0

Library of Congress card number
87-90573

Production by
Limousin World
Fort Collins, Colorado

Printing by
Jostens
Topeka, Kansas, USA

FOREWORD

The pages which follow tell the story of Limousin in America; how in 18 years these unknown French cattle moved to fifth place among the established beef breeds in the United States.

This is a phenomenal story and would not have happened but for a fortunate combination of essential ingredients—the right cattle, the right people and the right leadership.

The right cattle. Centuries of selection in France fixed genetic characteristics in the Limousin breed that fit the beef production process in the United States today.

The right people. The breeders who brought Limousin its place today were the real cattlemen and women who recognized the strength of the French breed, saw where modifications were needed to fit the American scene, and had the patience and faith to mold the American Limousin. These were men and women who really believed in the breed. Expansion of the breed was a worthy cause that inspired them to work to spread the good news about Limousin.

The right leadership. Through the years, the members have elected directors to the NALF board from across the country, with experience in every phase of meat production, from the cow/calf operation through the feedlot and even the butcher shop. They lived through the "breeding up" programs of the early years and the purebred and fullblood operations that exist today. These men and women made the decisions and established the policies through the years that were a necessary foundation for the breed's progress. They recognized that they were responsible to all of the breeders and for every aspect of the entire breed.

A combination of the wisdom of the directors and plain good fortune brought NALF three executive vice presidents, each with particular talents and background that fitted him for a certain phase in the introduction and expansion of the Limousin breed, and each has served us well.

Through all of this time, a unique leader for the breed was Dale Runnion. He took the risks along with the rest of us, kept the faith, and contributed his great enthusiasm and sound judgment. His broad experience permitted him to produce for us a breed magazine of the highest quality, which was one of the building blocks in the breed's success story.

As those of us who have come along this Limousin trail read the history of the past 18 years, we realize that it has never been easy. It was just something we had to do. Our satisfaction comes from knowing for certain that Limousin is here to stay because we have shown that the breed has an important contribution to make to the American beef cattle industry.

Floyd McGown

DEDICATION

Louis de Neuville
Chairman of International Limousin Council

To the pioneer breeders of the ELPA group in France and those 99 Founder members of the North American Limousin Foundation who helped introduce the breed into North America.

—Louis de Neuville

CONTENTS

For well over a hundred years Limousin cattle have been winning Grand Championship awards over all French and foreign breeds at the International Agricultural Show in Paris. This famous sire is "Conquerant", who won the Grand Prix d'Honneur in 1886—the year the Herd Book Limousin of France was organized. This artist's sketch portrays the best of show over all the breeds in Europe at that time.

—Copy photo from French archives by Dick Goff

The History of
LIMOUSIN
In North America

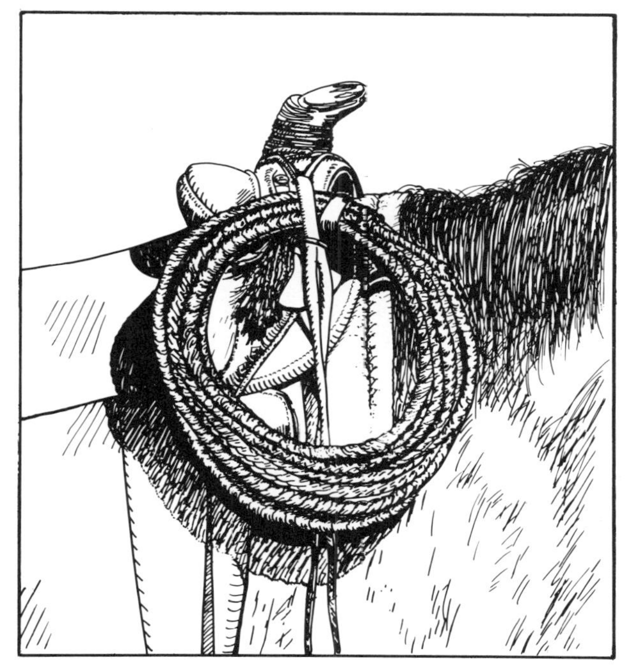

Part One 1967-1972
The Introduction of the Breed

by Richard Goff

Chapter 1
In the Beginning—Background on Cattle Imports

The first mention of the word "Limousin" in connection with a breed of cattle came to me from a friend. Dr. Harold Hill DVM had been head of the A.I. Bull Center at Colorado State University, but later had been working as a consultant for the Armour Beef Cattle Research Project in the Bahama Islands.

The Armour group was interested in working with Charolais cattle, but they wanted to import new bloodlines directly from France to the Bahamas. Dr. Hill was sent to France to select the first shipment of these animals and while there he visited the National Agricultural Show in Paris. It was there he saw his first Limousin cattle, and he was impressed.

On his return to Colorado some time later, Dr. Hill stopped in to see me. I was setting up the Pan-American Charolais Association, a group then being organized to register Charolais and Charolais-cross calves sired by artificial insemination.

At that time only dairy cattle breeds had approved A.I. calves for registry. Nearly all beef breed organizations were opposed to the practice. They felt, right or wrong, that it would ruin the bull market for their breed.

As events proved later, artificial insemination did not ruin the bull market but it did change values considerably. The top performance herd sires began to benefit from semen sales and their value soared.

Average bulls, on the other hand, were in less and less demand. More and more of them went to slaughter or were sold into commercial cow herds at prices that inspired a wave of cross-breeding ex-

Dick Goff liked what he heard about the big red Limousin cattle native to the Haute Vienne region of France. He conceived the founder-member concept for a multi-purpose association, and with the support of his many friends founded the North American Limousin Foundation. He served as the organization's first executive vice-president.

periments among the range men.

So when Harold Hill told me he had seen another breed of cattle at the Paris show that had great potential, I was interested but not greatly concerned. France seemed far away.

"What are they?" I asked.

"They are called Limousin," he replied, "and they may be able to give the Charolais a real run for their money."

"What do they look like?" I asked, curious to know more about any possible competitor, however remote.

"They are a beautiful golden red, slightly smaller than the Charolais, and they produce a terrific carcass," he answered.

At that time the possibility of another breed arriving on our shores from Europe seemed very unlikely. Even the Charolais then in the United States had come into this country from Mexico, all descendants of a herd of the white cattle that were imported to Mexico in the early 1920's. This was several years before a livestock quarantine treaty against European cattle was signed by Canada, the United States and Mexico.

This treaty was the result of a violent outbreak of Hoof and Mouth disease in Mexico in the 1940's. The possibility of this costly contagion sweeping the U.S. livestock industry had been cause for widespread alarm throughout the entire North American continent. The U.S. and Canada agreed to cover the staggering cost of eliminating the outbreak in Mexico, providing the three nations agreed to a continuing embargo against future livestock imports from any nation where Hoof and Mouth and Rinderpest diseases were known to exist.

Knowing this background, I was very complacent about this unknown French breed offering any real threat to the growth of Charolais, the "Silver Cattle with the Golden Future."

Little did I realize how familiar I would become with the provisions of that Tripartite Quarantine Treaty and those mysterious golden cattle of France over the next 25 years.

One thing the Charolais cattle had done for the beef industry in America was to demonstrate the cash value of crossbreeding to the commercial cattleman. At the same time, the development of new techniques for freezing and packaging bull semen had made it possible for this product to be processed and shipped easily and economically. Artificial insemination then became simple and practical. The result was a major new influence in cattle breeding.

It meant that an outstanding, high performance bull of any breed could now sire thousands of calves in all parts of the country from an infinite variety of cows. It meant also that a small breeder could "sample" another breed inexpensively by A.I. service to his own cows. The cost of a dozen ampules of semen from a high performance bull was relatively small, and the services of a capable A.I. technician allowed an experienced cowman to see how it was done.

The A.I. business then became the "wild card" in the breeding game and anybody could play. By refusing to register A.I. calves from purebred or commercial cows, the traditional breed organizations kept the lid on this game for a short time.

But the stakes were too high and the technique too simple to keep it bottled up very long. In the early days of A.I., semen was frozen into pellets and

Prince Pompadour NIM 1, by Baron was bred at Domaine de Pompadour in France. He is pictured here at 16 months of age at Rimouski, Quebec, just after release from quarantine, with Yves Caron, manager of the bull stud in Rimouski (left) and Charlie Moore of Coon Rapids, Iowa.

Adrien de Moustier, founder of Bov Import, imported the first Limousin bull (Prince Pompadour) to North America. He was active in the formation of both the North American Limousin Foundation and the Canadian Limousin Association.

these were carefully transported in alcohol and dry ice.

When liquid nitrogen and mass production in glass ampules appeared, another great leap forward took place. Insulated metal containers and air freight in jet aircraft appeared at about the same time to make shipping semen easy, safe and fast. Scientists had found that cattle, unlike humans, had almost infinite combinations of identifiable blood codes that could be used to identify not only a bull and a cow, but also their progeny.

The solution was there. One of the most prominent centers in this new scientific detective work was the Serology Laboratory of the University of California at Davis. The head of this renowned institution was Dr. Clyde Stormont; he became the guru of the blood takers and the Sherlock Holmes of the breed secretaries.

If the Charolais could do all these things in crossing with our great old Hereford, Angus and Shorthorn cows, there must be other breeds out there in other countries of the world. In addition, the Charolais cattle that were doing so well had a very narrow base. The "bull-seekers" needed new bloodlines, preferably direct from France. It was this possibility that had inspired the project by Armour in which Harold Hill was engaged.

Here the spectre of the iron-bound Tripartite Treaty that grew out of the Mexican Aftosa (Hoof and Mouth disease in Spanish) panic of the 1940's appeared to block that idea. The terms of the treaty gave the Animal Health Section of the U.S. Department of Agriculture broad veto rights on any program to import cattle from Europe by either Canada or Mexico.

Good purebred Charolais bulls were mighty expensive and the principle source of these purebred sires were breeders in Texas and the southern states who had been in on some of the few legally imported animals from Mexico. These established Charolais breeders were very protective about their lucrative bull market and they controlled the breed organization, the American International Charolais Association. They refused to register A.I. calves.

In my previous work with cattle publications and cattle associations I had met several cowmen of a "rebel" group from Montana, Wyoming and Colorado that decided to form their own association. These Charolais rebels appointed a committee and set about forming an association that would certify performance records and register or record A.I. calves in the United States, Canada and Mexico. One of them jokingly referred to the group as the ''Renegade Charolais Association.'' They came in to see me one day for some help in promoting their idea.

We decided to call it the Pan American Charolais Association and to also develop a working agreement with Charolais breeders in Mexico and Canada. At that time they had only 18 members, but we soon discovered their plan to record A.I. calves was the greatest membership promotion feature any beef breed association ever had.

I became executive secretary of the organization and wrote some articles for a number of livestock magazines. As the artificial insemination business expanded rapidly, so did we. We went from 18 members to 500 in two years and the A.I. recording program for crossbred calves opened up new market areas for both bulls and semen. At that time there was no Charolais association in Canada and several breeders there became members of our Pan American group in order to get their heifer calves recorded.

Later, when a Canadian Charolais Association was formed, they adopted many of the features we had developed and a reciprocal agreement was made to exchange registration certificates. Earlier we had made a similar agreement with the Herd Book Charolais of Mexico and their certificates were accepted by our organization.

In the meantime the American International Charolais Association had decided that artificial insemination was here to stay. The Pan American group worked out a merger plan with them and the two organizations were joined. If both were going to accept A.I. calves for registration, then the purpose of the original ''rebel'' group had been achieved and both outfits decided there was no point in having two groups competing for the same breed. As a result of the merger agreement I became promotion director for the combined operation.

Then, in 1964, the Charolais Herd Book of France was to celebrate their Centennial with an International Concours at Vichy. A number of Canadian breeders attended as well as some 55 Americans. For most of us it was the first opportunity to see the French cattle on their home turf.

This was how my wife, Jane, and I became tour directors for the first AICA tour to France, and how I accidently got my first glimpse of a Limousin.

Adrien de Moustier and I had become acquainted several years earlier through our mutual friend, Jerry Litton. Adrien's family had been raising Charolais since the time of Napoleon and he had been our house guest on several occasions when he visited the United States.

Adrien had volunteered to be our guide for the American tour group in the Nivernais country. He was sitting across the aisle in the big Mercedes autobus that afternoon during a tour of Charolais breeders in the area. He was aware of my interest in the Limousin breed and as we came roaring down a narrow country road about 50 kilometers southwest of Nevers, Adrien leaned over and pointed out the bus window.

''There,'' he said quietly (we were on a Charolais tour) ''are your Limousin cattle.''

I looked out over the field of wheat stubble to see a yoke of two golden-red cattle pulling a plow, with an aged French farmer gripping the handles and plodding along in the furrow.

Just that one brief glimpse! Then the bus careened over the crest of a hill and this fleeting tableau vanished from sight. It was such an incongruous scene, like something out of a medieval tapestry. There was no time for a quick photo, but the picture is still vivid in my memory.

My curiosity was aroused. I later asked beef experts at several state universities about Limousin, but at that time, no one seemed to know anything about them. Sometime after we had returned home, the *Western Livestock Journal* carried a brief article on the breed. It quoted a Canadian livestock expert who said they might possibly be the most efficient beef cattle in the world.

Then the international import picture cracked wide open! Early in 1965 Canada announced plans to establish a livestock import station that would allow breeding cattle to come into that country from Europe.

Dr. Kenneth Wells, veterinary general of Canada, and Mr. Harry Hays, minister of agriculture, had taken note of the ''tail-end'' position of the cattlemen in their country, and the American pressure to get new bloodlines into the U.S. They had seen the results of crossbreeding, and they decided that Canada could benefit from such an import program.

It was a brilliant economic and political move. Overnight, Canada became the center of interest to

This is Baron, sire of Prince Pompadour, at the Domaine de Pompadour in France as he looked in 1970. By this time he had become the grandsire of thousands of Canadian and American calves that were the base for several hundred foundation herds of Limousin-cross breeding programs. Photo by Dick Goff

the "breed-seekers" of North America. Canadian cattle breeders were now the key to all the new bloodlines and the new "exotic" breeds of the world.

Then the most startling part of the program was announced. Work had already begun on the station itself. The facility was to be located on Grosse Ile in the St. Lawrence River, a few miles downstream from the city of Quebec, and the station would be receiving animals that fall for a six-month quarantine period.

Grosse Ile had been the Canadian counterpart of Ellis Island for the U.S. during the years of the great immigration flood of people that came from Europe to the "new world." Like our Ellis Island quarantine facilities, however, Grosse Ile had been closed for more than half a century.

Now it was to be refurbished, laboratories installed and equipped, stalls fitted into old dormitory areas, feed storage facilities prepared, and it would be ready to receive cattle by the fall of 1965. The

news caused a sensation in the livestock industry.

But what about the USDA position on the project? Like most international agreements there was an "out" provision in the three-nation quarantine treaty. This allowed the other nations to import cattle, providing they could maintain the stiff sanitary regulations set up by the Animal Health Section of the U.S. government, and meet USDA approval.

For many years the United States had maintained an animal quarantine station on Plum Island in the New York harbor area. This station was used for the importation of livestock from the countries that were considered "disease free"—primarily England, Scotland, Ireland and a few other remote areas.

The Canadian government adopted the same testing and quarantine procedures which the U.S. had developed for Plum Island. In addition, they invited the U.S. government to send an experienced animal health veterinarian to monitor the Grosse Ile operation as a live-in observer of the testing and laboratory procedures. In this way the animals coming through the Canadian quarantine should meet all the requirements of animals entering the U.S. by way of Plum Island. And, since it was obvious that a major market for these imported animals and their progeny would be to American breeders, this procedure should open the border to U.S. buyers. As we found out later, however, nothing was ever quite that simple where our government bureaucracy was involved.

The Canadians did a remarkable job of getting their project under way and the first animals arrived by boat from Brest, the French export quarantine center, in the fall of 1965.

By May 4, 1966, the first cattle had completed the quarantine program and were released to their Canadian owners for a further three-month quarantine on the farm. This meant they could not be sold, nor could semen be sold from the bulls during this 90-day "observation" period. And, they were to be kept isolated from other animals on the same farm. However, prospective "pardners" could see the cattle during this time.

Furthermore, the Canadian government announced the terms of the import permits specified that none of the imported animals could be taken out of the country for a period of five years from date of entry. It was another master stroke of economic policy that insured the future position of Canada as the seedstock center of North America, at least for the new Continental breeds.

The first year's imports were entirely Charolais, mostly bulls. In all there were about 100 head, although the facility at Grosse Ile was set up to care for about 250 animals. At this time the demand for Charolais was at its peak and American breeders, many of them well supplied with cash, made a hurried pilgrimage to the new mecca of the cattle industry. It was a bonanza to the Canadian cattlemen and they made the most of it.

The average cost of an imported French bull, delivered to the new owner's farm in Alberta, came to about $7,500. Several "half interest" deals were made with the eager Americans for as much as $25,000 U.S. One Canadian wrote me offering a one-eighth interest in his bull for $7,000 U.S.

It was a profitable deal for the fortunate Canadians who were lucky enough to draw an import permit. Many of these breeders were small operators to begin with, but one permit could put them in the purebred business very quickly. The story was told that several of the applications for the first importation were simply written on postcards.

When the 1966 permits were issued in the early spring for animals to be released in 1967, one breeder had applied for a permit to import a Pie Rouge bull, the rugged red and white French version of the Swiss Simmental. This began the introduction in the next few years of several entirely new beef breeds into the North American livestock world that had never been available before.

Previously, Adrien had made connections with a French agricultural school in Quebec province, about 150 miles down the St. Lawrence from the city of Quebec. With their help he had obtained permits and had imported two of his own Charolais bulls in the

Big strong calves like these delighted both the breeder and commercial man and helped move the breed into a more prominent position for the future.

fall of 1966. I flew to Quebec and met him there when these animals were released early in 1967. Although neither of us knew it at the time, the first Limousin bull calf to set foot on the North American continent was then enjoying the taste of his mother's milk in France.

Earlier in the day we had stood on the snow covered pier at Port Levis and watched his Charolais bulls unloaded from the river freighter from Grosse Ile into a truck, the last leg of their journey to a new home in Canada.

That evening we went back to the historic Chateau Frontenac and talked over the possibilities of a Limousin project. Adrien said the French-Canadian officials of the school at Rimouski thought they might get him three permits that spring for import animals from France. If this happened, he promised that one of them would be a Limousin bull, providing he could find a worthy animal at a reasonable price. I agreed with him the financial risk was certainly greater for an unknown breed, but we had seen the first Simmental on that boat. I told him I thought this was the time for Limousin; we might never have such an opportunity again. The next day I flew back to Denver and began thinking about promoting an entirely new breed in Canada and the United States.

By this time I was no longer involved with the Charolais association. My advertising business had grown and I had merged with an old friend, Don Clair, who specialized in industrial advertising. I had also acquired the Franklin Laboratories advertising account. They were a respected producer of livestock vaccines and pharmaceuticals. Dr. Franklin had developed the first commercial vaccine for Blackleg nearly half a century before. It was the largest advertising account of this type in the region and they were wonderful people to work with. Furthermore, they had an extensive marketing organization in Canada and Mexico, as well as the western half of the United States.

My experience with the old Pan American Charolais Association had been an ideal training course for this new Limousin project. It was familiar territory and I hoped I could plan well enough to avoid some of the usual headaches of a new organization.

Some time later, Adrien called from Paris. Only one permit had been granted. Now he had a tough decision to make, and an expensive one. What bull? What breed?

By this time, over 300 Charolais bulls had come into Canada from France. This was enough to supply a lot of semen to the American market. In addition, the first Simmental semen was just becoming available.

"Adrien," I told him, "there are lots of Charolais in the U.S. and Canada, but there is not one Limousin on this entire continent. Why don't you go down to the Limousin country and look at them?"

On August 21, 1967, Adrien wrote me a letter. "Just came back to Paris to find your letter. I have to tell you that I have taken a decision concerning the Limousin, and I have bought the best calf available in the breed...from the French government farm! I have seen outstanding cattle, and I will tell you more about them.

"The breed (association) is poorly organized, except for a small group of breeders who have developed a tremendous organization." (This was ELPA, headed by Louis de Neuville.)

"I am sure now," he continued, "the Limousin will have a big appeal in your country. The bull is being tested this week (for export), and I will let you know more," he added.

One of the problems Adrien had encountered on this first visit to the Limousin country was to find unvaccinated bull calves that would be eligible to enter the French export testing program. It had become the practice among the Limousin breeders to vaccinate their calves against l'Afteuse (Hoof and Mouth disease in French) in early calfhood. These vaccinated calves then developed anti-bodies against the disease. If subjected to the tests used in the export quarantine program, these animals would show a positive reaction, just as any animal would that had the infection itself. This meant the animal was rejected for export.

It turned out that ELPA (Eleveurs Limousin de Plein Air), the organization that Adrien had mentioned in his letter so favorably, came to his rescue. They were a group of some 20 or more breeders who

Dick Goff and Louis de Neuville combined their efforts with Adrien de Moustier to start the Foundation, and find and import the first Limousin bull to North America. They are shown here at a recent National Western Stock Show.

had adopted a rigid program of herd improvement. This included performance testing, rigid culling of replacement heifer candidates and a practice of "open air" calving that was contrary to cattle industry custom in France and most of Europe.

When Adrien announced he was looking for the first Limousin bull for export to North America, everyone in ELPA was interested to see that he found a good one. Since the French law required calves be vaccinated before weaning, many breeders had already immunized their calves, even though weaning might not take place until late in September or early October.

Then it developed that an outstanding bull calf at the Domaine de Pompadour was unvaccinated, but he had been sold! The Pompadour operation was a world renowned breeding research center owned by the French government. Devoted to Limousin cattle and to the Anglo-Arabian horses bred in France for steeple-chase racing, it was under the direction of M. Pechdo, who was also an ELPA member. The buyer of the bull was Emile Chastanet, one of the outstanding Limousin breeders and also an ELPA member. All felt it was vitally important that the first Limousin bull to step ashore in Canada be a top quality representative of the breed. Louis de Neuville, as president of ELPA, got everyone together and Chastanet agreed to give up the bull. As Adrien wrote me, he was the "top of the breed" in that year's calf crop.

From Limoges, where the bull went through preliminary tests, the animal went to the export quarantine station at the French seaport of Brest. Here he joined the other candidates of other breeds for a 30-day test program. If all the cattle passed, they were then loaded aboard a freighter (still in quarantine) for delivery to Grosse Ile, and they entered the intensive six-month quarantine program by Canadian authorities.

One of my accounts was a new artificial insemination center near Denver and I became familiar with the processing and marketing of beef bull semen. Until this time nearly all the commercial activity in frozen semen had been concerned with dairy cows. Here the cash return from breeding to bulls with a record of top milk production was well known. In beef cattle, the work of finding bloodline values was just beginning to show results. The spread of commercial crossbreeding had also increased interest in beef performance research.

In my search for information that might be of value to my advertising accounts, I met Dr. Richard T. Clark, then head of the Western Beef Cattle Research program of the U.S. Department of Agriculture. Headquartered in Denver, he had under his direction the Range Cattle Research Center at Miles City, Montana. Here the government had spent more than 20 years in developing the celebrated Line One strain of high performance Herefords.

As the interest in crossbreeding grew, Dr. Clark decided they should investigate the use of Charolais in a crossbreeding program with these superior Hereford cattle. As a result I became well acquainted with this remarkable man. Born in Scotland and educated there, he came first to Canada and then to the United States.

I made several trips to Miles City with Dr. Clark after this crossbreeding project got under way. His comments, as we looked at the first crossbred calf crop, and his wise observations on the subject of cattle breeds in general, gave me a background that had a major influence several years later on the basic plan for the North American Limousin Foundation. This despite the fact that he died of a heart attack two years before the NALF program was conceived.

One comment of his, in particular, had impressed me on a trip back to Montana to see the second crop of the Charolais-Hereford calves.

"We have produced a greater improvement in weaning and yearling weights by two years of crossbreeding than we have in 20 years of selective line breeding," he declared one evening.

"And yet, when it comes to crossbreeding," he continued, "you must always keep in mind there is a greater difference between individual animals within a breed than there is between the breeds themselves." I saw this demonstrated many times during the next 20 years.

From my own experience too, I knew there were basic problems that any new breed association would encounter. The modest membership fees we had set for the Pan American organization had been inadequate to set up a recording system that would accommodate the workload as registration applications came in.

We had expected registrations to follow memberships by about a year, but it didn't work out that way. The pragmatic cattlemen wanted to see calves on the ground before they decided to join the new outfit, and nobody could blame them. As a result,

In 1972 the NALF board was filled to 15 members and was pictured after the July convention. Seated left to right: Floyd McGown, Fred DeMier, Cadet Oxandaburu, James Easland, Stephen Garst, John Moore, Robert Vantrease, executive vice president. Standing left to right: James L. Baldridge, David T. Allard, Bryant Harris, H.M. Jordan. Charles L. Moore, Burwell Bates, president, R.D. Bennett and Walter Shatto. J.V. Elliott not available for the picture.

we often received a membership application along with a bundle of A.I. calving records. We were never able to catch up.

And there was always the vital question of front money. A lot of work had to be done long before the first cow was bred. Adrien had agreed the Limousin organization would have exclusive sales rights to all the semen from the bull, if he produced! So I felt the first semen available should be restricted to members only. What would this "priority package" be worth? At this stage it was anybody's guess.

As the winter passed, I talked to more and more stockmen. Some were skeptical of such a nebulous proposition. Hardly anyone in the United States had ever seen a Limousin bull, for that matter. That included me, of course.

There were two basic approaches I could take that would interest most of these people: 1-To form a corporation and sell stock in the project; 2-To form a combined breed association and research foundation with an "equal opportunity" deal for each member, a much more idealistic approach.

Time was passing and the proposal had to be writ-

ten. After several different plans had been outlined and rejected, I decided on a package deal that would offer a "three-way bundle" for each new initial member in the "North American Limousin Foundation." This would be incorporated as a registration, research and promotion entity for the breed. It would have broad powers to operate in Canada, the United States and Mexico. It was quite honestly modeled after many of the aims of the Pan American Charolais Association that we had developed and proved nearly 10 years before...with one major exception.

We had been assured of a monopoly on the only Limousin bull semen on the continent, and this would be delivered on a priority basis to the first "Founder" members who would also underwrite the beginning program. Each Founder Member would put up $2,500 for this special membership, an estimated 30 ampules of semen and a priority on additional semen as long as we had some control of the supply.

Each member would participate in the initial research and development of the breed, and these

16

founders would then vote on the opening up of a regular membership at $100. These memberships would carry no priority on semen, however. Adrien had already applied for two more bull permits for the coming year and we knew other Canadians were looking at Limousin.

In the meantime, on an ice-bound Canadian island in the lower St. Lawrence River, Prince Pompadour, the standard bearer of this new breed in America, was undergoing weeks of testing. Not until spring came to break up the ice jam that locked Grosse Ile away from the world, would he be free to sample the green grass of a new continent.

Whenever the Prince Pompadour NIM 1 turns up on an extended pedigree today, it is a special kind of tribute to Adrien de Moustier who gambled a good many thousand francs on the project. And to those dedicated members of ELPA, to M. Emile Chastanet, M. Pechdo and to M. Louis de Neuville, who rallied to help him find the best kind of bull to break the new trail to North America.

In the French Herd Book he was Castor, son of Baron out of the great Pompadour dam Uxoria, and a grandson of the renowned bull, Laureat, who had produced some of the outstanding performance sires of the ELPA Group. When the bull left France, he was Castor, and I didn't like the name from a promotion standpoint.

Tommy Thompson and Leland Dudley, Hampton, Iowa were Founder Member # 21. The Dudley-Thompson partnership held the first Limousin sale in North America, consigned the top halfblood heifer at the first National Sale and provided strong leadership in the state and national organizations.

When Adrien and I talked the next time on the telephone I told him about this concern. Since the calf was born on the historic Pompadour estate and made famous by the glamorous Marquise de Pompadour, mistress to Louis XV, I suggested we tie that touch of glamour into his title. "Let's call him Prince Pompadour!"

"Great," Adrien laughed, "I like it."

Later we found that M. Pechdo and Louis de Neuville were also pleased that the young vanguard of the breed would carry the banner of their historic research center.

Meanwhile I had been talking to a lot of old friends in the cattle business. One long time friend whose judgement on the cattle industry I respected greatly was Forrest Bassford, editor of the *Western Livestock Journal.* Forrest and I were sitting together on a Western Airlines flight to Casper one morning on our way to attend Bob Purdy's Charolais production sale at the famed 28 Ranch near Buffalo, Wyoming.

I outlined my alternatives to Forrest, adding that I didn't want to make the "Founder's" ante too high, but I didn't want to handicap the program by running out of funds at the first crisis that came up. I added that I had made projections on any amount from $1,000 to $5,000 for the first few priority members.

Forrest was intrigued with the program I had in mind. I said I knew I could make a lot of money for a few people, or spread it out a bit and build enough wide general interest in the breed to gather considerable momentum for the long haul.

"How much is price going to be a factor in lining up the first members or stockholders?" he asked. I explained that some of the people I knew would put up almost any amount if they were assured control of the breed for the first few years. On the other hand, I pointed out, it would help the long range development of the breed if I were able to attract some good solid commercial cattlemen to come in for the long haul. I knew both kinds of people, and the key would be the initial presentation of the program. Once launched it would be difficult to change.

We discussed the pros and cons of the various propositions until the plane landed at Casper, and we rented a car to drive from the airport to the 28 Ranch.

After the sale was over the next day, I talked to

Bob about the proposal. I admired Bob's business acumen. We had been through the "Charolais wars" together and I had known him for more than 10 years when the Limousin idea developed. He had an MBA from Stanford and a CPA degree from Wisconsin. He had taken over the family business when his father became ill and turned it into an industrial empire. When he found that making more money seemed pointless, he came to Wyoming and bought the historic 28 Ranch and the adjoining Billy Creek operation.

Starting with Angus, he had then begun experimenting with Charolais. He had been one of the half dozen "renegades" who had started the Pan American project. No one I knew had a better understanding of business organization and promotion potential.

By this time, however, he was feeling the effects of a developing problem with diabetes. I could see he was losing some of his enthusiasm for a new challenge, but the idea was intriguing, providing the cattle could live up to our expectations. When I left the next day, he grinned and said that whatever I decided to do, to count him in as a Founder.

Chapter 2
Forming the Foundation

Spring was coming. Before long our first Limousin would be released from his ice-bound quarantine on Grosse Ile and be the first of his race to walk the pastures of this continent.

To handle his progeny records we needed an organization that would combine breed registration and pedigree service with a performance recording and research entity as well. If the society or association were to incorporate all these features, the word "Foundation" might be a more fitting title. It would distinguish our organization from the older breed associations and might also give us some future promotion advantages. That is why it became the North American Limousin Foundation.

I called Willis Carpenter, a young attorney who had just opened his own office in Denver, and told him what I had in mind. He was the son of Farrington R. Carpenter, one of the West's best known Hereford breeders, as well as the first administrator of the Taylor Grazing Act which was the predecessor of the Bureau of Land Management, and one of the principal organizers of Performance Registry International.

When we got together the next day I asked him to prepare articles of incorporation and by-laws that would let the Foundation do almost anything the officers deemed worthwhile to develop the breed and advance the beef cattle industry. In addition, it should be chartered to operate on an international basis, for I could see that we might be recording cattle and doing research in other countries.

About this time several things happened. Charles L. Moore called me from the Denver airport. He was enroute from Phoenix, Arizona to his home in Iowa and had a stop-over and plane change in Denver. Charlie was on the Board of Directors of the American International Charolais Association and General Manager and partner of Star Charolais Farms, one of the most prominent purebred herds in the Midwest.

Charlie had just attended an AICA directors meeting in Phoenix. It had been one of those affairs in which differences of opinion had caused a lot of dissension. Charlie was disgusted and declared he was a cattle breeder and not a politician.

"I'm about to the point where I'd get out of this Charolais business, if I knew of another breed anywhere near as good," he muttered.

"Charlie," I laughed, "stay right where you are and I'll come out and have dinner with you. I've got a project that might interest you."

That evening at the airport restaurant I told Charlie about the Limousin proposition, the fact that the bull was then in quarantine and would be released in a few weeks, plus the outline of a Foundation that I hoped would open up some new areas for breed promotion and development work.

When I went home later that night I promised to keep him posted on developments. He said he would think it over and let me know.

18

"Frankly, Dick," he said finally, "it sounds mighty interesting."

Thus, Charles Moore got the first "cold turkey" sales pitch on the Limousin Foundation idea. We had worked together when I was promoting Charolais cattle, and I had a great deal of respect for his judgement and his ability as a cattle breeder. His questions, comments and suggestions that evening gave me a lot of encouragement.

A few days later I saw Bruce Waddle at Longmont, Colorado. Bruce was a successful businessman and a Charolais breeder also. He and his wife Beverly were a great team. They had started out with a small engraving business and built it into a very successful precision metal processing plant. Their Charolais herd had developed about the same way. It started small and developed into a prestige herd. Their top herd sire, Poker Chip, had become nationally famous. I was interested to get their ideas on the Limousin project.

After outlining the story and the plan for setting up the Limousin Foundation, I asked for comments. They too were intrigued with the idea, but Bruce felt he was just now making progress with his Charolais and knew that starting over again with a new and unknown breed was a long range program. I agreed but suggested he could start experimenting with Limousin semen on some of his percentage Charolais cows and see what they thought of the results. I said I would keep them informed and give them a report as soon as I got back from Canada.

Adrien de Moustier called that morning and informed me the bull would be released from quarantine the next week. I promised to meet him in Montreal and we would drive to Port Levis, across from the city of Quebec, to see the historic unloading of North America's first Limousin bull.

By the time we arrived dockside at Port Levis a few days later, the unloading was already under way. The animals came out one at a time, were loaded into a crate and hoisted out by the ship's crane and put down into the holding pen and truck loading ramp. It was slow, but the animals were handled as gently as a crate of eggs.

We went aboard the freighter to peer down into the hold to see if we could spot our bull. Everything so far had been white Charolais. As we looked into the open hatch we could see a mass of white Charolais with one lone red animal right in the mid-

Bruce Waddle, F. M. # 3, is still one of the leading Limousin breeders in the United States. He presently serves on the NALF board and is president of the Colorado Limousin Association. Son Brad, pictured at right, manages the present Waddle herd.

dle. There was our Limousin! Seen from above, he looked long and slender in comparison to the heavier Charolais bulls. But what a beautiful red color; a different shade from the familiar Hereford red. We didn't get much chance to admire him for the van was waiting to take him to Rimouski and he was loaded into the covered truck immediately. Technically, he was still in quarantine, so we would have to wait until we got to Rimouski to get a good look at the animal. There he would be under farm quarantine for another 60 days, but we could see him and take a few pictures.

The next morning we drove out from the St. Louis Hotel in Rimouski to the small farm where he would be secluded for this extra two months. The farm was on a high ridge that faced the broad waters of the St. Lawrence, about 150 miles downstream from Quebec. At this point the river was 32 miles wide and the north shore was barely visible that sunny spring morning. Our young bull was in a small holding pasture, looking around at this new world and enjoying the sun and the view. His head was up, his eyes bright and flashing and the sunlight glistened on that gold-red coat. He looked every inch the Prince of Pompadour.

"He's magnificent!" I told Adrien.

We went into the pasture and walked around the bull several times, admiring him from every angle. The bull seemed as curious about us as we were about him. He stood quietly, watching us in a bemused manner, it seemed to me. He had a cavalier air that inferred our plaudits were expected for one of his illustrious breeding.

I had seen a lot of young bulls in my time, but this was the most exciting animal I had ever laid eyes on. Part of it was the anticipation, of course, and part the spectacular setting with the glint of sunshine on that unusual hair coat. From that moment I felt the future of Limousin in America was assured.

Adrien and I drove back to the hotel and began to map out a plan of action. He had to go back to Montreal on business and I wanted to make arrangements for a semen test and a thorough physical check-up on the animal. We had a lot riding on that handsome yearling.

After Adrien left I got on the telephone to Dr. Jim Scott, a veterinarian who specialized in bull fertility work. He had an office in Fort Collins, Colorado and also worked at the Bull Center operated by Colorado State University there. Jim and I had discussed the Limousin project earlier, so when I called to tell him about the bull he was interested and agreed to fly to Rimouski with enough portable equipment to run the required tests.

The next evening Dr. Jim arrived at the Mont Joli airport near Rimouski. We still had a couple of hours of daylight left so I drove Jim directly out to see the bull, the first order of importance.

Jim was also impressed with the young bull. It was about the first of May and the small pasture he had for an exercise area was covered with dandelions. He stood there in the evening sun on what appeared to be a bright golden carpet. It was a striking picture for America's premier Limousin herd sire. He looked like he could handle the job.

Early the next morning we were there with Jim's equipment to run our first field test on the young bull. Jim examined the semen under a high power microscope he had brought with him and, from a grid count, estimated that he produced about a billion, four hundred million sperm cells per cubic centimeter. Quality and motility were excellent, he concluded, adding that this one collection would have produced about 150 ampules of commercial semen. Unfortunately, Canadian quarantine rules prevented us from transporting raw semen off the farm, so the possible 100-plus calves that could have resulted were lost to posterity.

I shot a number of photos of the bull, now officially christened "Prince Pompadour." That after-

noon we caught a plane back to Denver.

A few days later, on the 6th of May, 1968, we had a preliminary meeting of a handful of people who had indicated an interest in this new-to-America breed of cattle. By this time, Willis Carpenter had prepared incorporation papers for the North American Limousin Foundation.

At this first meeting we outlined a program that would utilize the accumulated genetic and performance information of these animals for future use in breeding programs, starting from Bull Number 1. It was an unusual opportunity, if we could get it started properly.

There were nine people at the opening of that initial meeting to discuss Limousin cattle and the feasibility of the Foundation plan. They were: Robert H. Purdy, Buffalo, Wyoming; Charles L. Moore, Coon Rapids, Iowa; Bruce Waddle, Longmont, Colorado; Dr. James Scott, DVM, Fort Collins, Colorado; Adrien de Moustier, Paris, France; Richard M. Patterson, Denver, Colorado; Willis Carpenter, our attorney; William W. Smutz, Denver, who had developed an advanced computer

Bob Purdy, Buffalo, Wyoming cattleman, businessman, served as president of the North American Limousin Foundation for the first three years.

program for his livestock management service; and myself. Adrien de Moustier was there to discuss the bull and the Limousin breed in general. Sherman Ewing of Alberta, Canada, arrived later due to a delayed airline schedule, bringing the final total to ten.

As acting chairman, I began the meeting with some background information on the Limousin breed, outlined the basic objectives of the proposed Foundation and introduced Adrien de Moustier as the owner and importer of Prince Pompadour.

Adrien discussed the Limousin in France and compared them to Charolais cattle, adding that he had already imported two Charolais bulls to Canada the previous year but felt the Limousin had some qualities which would make them extremely popular in this country. He had taken a few color slides of the Limousin, including the calf selected for export to Canada. He announced the bull was now on the farm in Quebec, near the town of Rimouski, and suggested the initial group might want to go there to see the animal.

Our attorney, Mr. Carpenter, then outlined the legal steps involved in setting up the Foundation and suggested we form an Ad Hoc committee to proceed with the organization work.

The group approved the basic plan of the $2,500 Founder Memberships and authorized the committee and the attorney to develop a contract agreement with Adrien for the bull semen that could be submitted to our first elected Board of Directors who would be named at the next meeting.

Bob Purdy then asked who would be Member Number 1? I grinned and said I supposed it would be the first person to put his check on the table. Instantly both Bob and Charlie Moore slapped their checks on the conference table. Both were made out to the North American Limousin Foundation. Since all agreed it was an obvious tie, we decided to flip a coin and Bob won the toss to become Founder member number 1 and Charlie was Founder member number 2. Immediately Bruce Waddle handed over his check for number 3, Sherm Ewing took number 4 and Jim Scott became number 5. Of the five, Jim was the only one who had seen a Limousin.

With that impetus to membership we adjourned to see Bill Smutz's computer operation at his office across the street. This was long before the time of personal computers and Bill's equipment was impressive. It consisted of several shoulder high cabinets with whirling spools of tape, a keyboard console and a printer pouring out apparently endless folds of paper.

I told Bob Purdy this outfit looked like a paper manufacturer's dream machine, knowing that he owned interests in some paper companies in Wisconsin. He laughed and said he hoped this project would pay out one way or the other.

Bill Smutz had developed some computer programs for ranch and livestock management, a service that was still new in agriculture. One of his clients was Sherm Ewing, and it was primarily because of Bill that Sherm attended our first meeting. We all agreed Bill should work out the initial recording program for our new Foundation.

Another meeting was set for May 17 to review further details of the organization. These were basically pre-organization meetings our attorney felt were needed to get the new corporation under way. By this time, we had received membership checks for 10 Founder members as specified by counsel to meet the legal requirements for an official organization meeting. Accordingly, June 21, 1968 was set for this date and letters sent out to all who had indicated interest in the breed.

This official "organization meeting" brought out about 20 people with the official paid membership count of 12 at the beginning of the session. Those present at this historic meeting were: Robert H. Purdy, Buffalo, Wyoming; H. A. McCoy, Topeka, Kansas; Mr. and Mrs. Ben Price, Jr., Reading, Kansas; Dr. E. E. Wedman, Ames, Iowa; Forrest Bassford, Denver, Colorado; W. P. Hinman, Yampa, Colorado; Col. Edward Geesen, Agate, Colorado; Henry Gardiner, Ashland, Kansas; F. Wallace Gage, Longmont, Colorado; Charles L. Moore, Coon Rapids, Iowa; Dr. and Mrs. James A. Scott, Fort Collins, Colorado; Cecil Hellbusch, Denver, Colorado; Don A. Pavel, Fort Collins, Colorado; Bruce Waddle, Longmont, Colorado; Richard Goff, Denver and W. W. Smutz, also of Denver.

Since we were still meeting under the Ad Hoc organizing committee, I called the meeting to order and outlined the purpose of the proposed Foundation, the background of the organizational work accomplished at the other two meetings and explained that this meeting was to fulfill the requirements outlined by our attorney for creating a legal, non-profit organization.

For the benefit of those who had not been brought up to date on the status of the Limousin bull now in Canada, I described the situation there and the fact that we had a verbal commitment for all of the semen from this bull, with a contract prepared for acceptance as soon as officers were elected and the organization became officially in operation.

Dr. Jim Scott gave a report on his appraisal of the bull and declared his field tests showed excellent semen production and quality. In his opinion, the bull was an excellent choice for the initial herd sire for the breed in America. There was some additional discussion on the historical background of the Limousin cattle in France and prospects for the future of the breed. More details of the proposed nature and purpose of the Foundation was outlined to the group.

We had excellent news coverage on the event from a press release I put out the day after our meeting.

One angle of the new organization seemed to interest most editors; the fact the initial Founder members would be limited to 100 people who would invest $2,500 for semen, research and breed development work. It was a program that was unique in the annals of the cattle industry.

Our news release tells the story most effectively after all these intervening years.

NEWS RELEASE 22 June 1968
NORTH AMERICAN LIMOUSIN
FOUNDATION

DENVER—An international beef cattle research project was launched here last Friday with the formation of the North American Limousin Foundation. It marks the first introduction of the French Limousin beef breed to the North American continent.

Prince Pompadour (in his mature state) left a fitting legacy of an estimated 60,000 offspring in the United States and Canada. His daughters formed the basis for many foundation herds in North America.

The group will initiate an extensive cross-breeding program with semen from a Limousin bull imported into Canada from France last fall. As many as 10,000 head of calves from this one bull may be involved in the plan to gather crossbreeding information, performance and carcass data over the next two years, the organizers said. Starting with the first calf, all of the genetic information, growth and carcass data will be fed into a computer for continuing statistical analysis, the group revealed.

Officers of the new organization are: President Robert H. Purdy, Buffalo, Wyoming; Vice President Sherman Ewing, Claresholm, Alberta, Canada; Secretary-Treasurer W.W. Smutz, Jr. and Executive Vice President Richard Goff, both of Denver.

Elected to the Board of Directors, in addition to the three officers, were: Stephen Garst and Charles L. Moore, both of Coon Rapids, Iowa; H. A. McCoy of Topeka, Kansas; Ben Price, Reading, Kansas; Bruce Waddle, Longmont, Colorado; and Dr. James M. Scott (DVM), Longmont, Colorado.

The Limousin is a French beef breed that has interested livestock geneticists because of its excellent carcass quality and marbling ability. Reports made at the meeting last week from French research data said it is a fast-growing, fast-maturing animal that is heavily muscled with an exceptionally high cut-out ratio.

Light red and tan in color, the animals are entirely new to this continent, although it is an old European breed and the French Herd Book has been registering bloodlines for nearly a century, Goff said. In France it is known as a "plein air" or "outdoor" breed that is carried outside the year around. An additional feature, according to French reports, is an almost complete freedom from calving problems. When born, the calves are relatively small and long and slender in shape. Breed officials in France say the practice of pulling calves is practically unknown in Limousin country.

The first Limousin bull to be used in the initial phase of the new project is a purebred animal especially selected by ELPA, a French performance testing and breeding organization. The bull was imported to Canada by Bov Import, Inc., a Canadian group headed by Adrien de Moustier, a prominent French cattle breeder and industrialist who is well acquainted in the United States and Canada.

The bull will be housed in a new beef stud and A.I. center, now under construction by the School of Agriculture at Rimouski, Quebec, a century-old institution operated by a Catholic church organization there.

Initially, the North American Limousin Foundation will be limited to 100 founder members who will contribute $2,500 each for semen, research and breed development work. Additional Limousin animals are scheduled to come into Canada this year, the officers said.

This project marks the first time any new breed of cattle has been introduced into this continent under such a far-reaching program of research and development, Goff pointed out. Subscribers to the Foundation include cattlemen from both Canada and the United States, with some prominent Mexican breeders also interested. Literature is now available in English, French and Spanish, it was announced.

We were now in business, and some of the outstanding livestock operators and research workers in the beef cattle industry were on our membership roster. In a matter of months we had raised nearly $100,000 to get the organization and its major program under way. We signed a contract for 95 percent of Prince Pompadour's semen production with Adrien de Moustier.

Now all we had to do was get the semen processed and delivered to the United States. This involved dealing with two governments, two departments of agriculture, customs and export officials, health regulations and international breed politics in general.

The Canadian government was very concerned about everything that was shipped out of Canada and the U.S. government was very concerned about everything shipped into this country, especially anything to do with the so-called "exotic" breeds of cattle. Our American officials sometimes gave me the impression the "exotics" were something they considered highly undesirable for this country, and we were downright unpatriotic and possibly a bit subversive to import French bull semen that would defile our American cow herds.

Adrien and I had to go back to Rimouski to see

about the business of getting semen shipments started as soon as possible. At our meeting I had urged as many of our new members as possible to come to Quebec with us and see the crown prince himself. As a long yearling, Prince Pompadour was a very showy animal and I felt the more people who got a firsthand look at him, the more enthusiasm we could generate.

Bob Purdy, Bruce and Beverly Waddle and Charles Moore made the trip with us. As I recall, it was around the last of July. Everyone was impressed with this new kind of beef animal. Even Charlie Moore, who had probably seen and judged more beef cattle than all the rest of us put together, had a big grin on his face as he walked around the bull, then ran his hands along the bull's loin muscle and over his hind quarter.

Charlie looked over at me, still grinning.

"I believe he'll do, Dick," he said, then added as an afterthought, "Of course we'll have to see his calves before we really know."

Bev and Bruce Waddle looked at him with their eyes shining. "He's beautiful!" Bev exclaimed, and that about summed up everyone's reaction for a first impression.

"How soon can we get semen?" Bruce asked.

That was the big question. The issue became more and more complicated as each day went by. First, the bull had to be collected at Rimouski, the semen diluted and shipped in ice to the government A.I. center at St. Hyacinthe, Quebec, a distance of 150 miles. There it would be processed, sealed in 1cc glass ampules and frozen in liquid nitrogen at roughly 300 degrees below zero (F). At that point it would be ready to be shipped in liquid nitrogen to Denver, we hoped.

But it wasn't that simple. First we had to get export papers from the Canadian government, releasing the shipment. Then we had to have import papers from the U.S. government allowing us to receive this "live animal product," accompanied by suitable health papers certifying the semen was free from disease of any kind.

In addition, we had to have further clearance from U.S. Customs authorities with a declared valuation and pay duty on each shipment. Each customs official had to open the nitrogen filled container and see each cane of ampules. That meant loss of nitrogen, exposure to light and any number of delays that would kill the sensitive sperm cells. It was the old battle with bureaucracy that cost us time and money for the next two or three years.

That summer of 1968 was critical, however. Every one of our members had cows to breed. Many had been holding some of their best females open awaiting the first shipment. There were more and more delays, and it seemed I spent a good part of every day on the telephone to Washington or Ottawa, or some lesser bureau on the border. Finally to cap it all, we received a letter from the USDA Animal Health Division in Washington postponing our first shipment for another 30 days in order to clarify certain quarantine requirements. We never found out why. It just seemed that someone, somewhere, was determined to make as much trouble as possible.

Finally, we did get a shipment into the Denver airport in August. There was a delegation of members waiting with their own nitrogen jugs to get that first allocation of six ampules. Colonel Geesen and his son Mick already had a "hot" cow in their breeding chute at their Agate, Colorado, ranch, about 50 miles east of the Denver airport. Bud Prosser was there representing International Beef Breeders, and after we had transferred that first cane of Limousin semen into the Geesens' nitrogen jug, he took the balance of the shipment to the IBB Artificial Insemination Center. The rest of us took off for a breakneck run to the ranch to see the first American cow bred to a French Limousin bull.

Present at this historic inception were Col. Ed. Geesen, Mick Geesen, Mick's wife Ginny, Bill Smutz, Dee Sterbenz, my secretary, Mark Curry, the Geesen ranch foreman and myself. To me it was the culmination of nearly four years of planning, selling and international "politics."

Two years later, in my first "Report to the Limousin," at Limoges, France, I showed the French breeders a color slide presentation of first- and second-cross calves from the Geesen herd that gave them their first visual appraisal of the hybrid breeding potential in the Limousin animals. They all seemed astonished at the power of the Limousin genes to influence the American commercial cow herds.

"You get SO much Limousin in your first and second crosses!" they exclaimed. Part of it was the fact the Geesen cattle were an ideal set of cows for such a project. They were a performance selected group

Colonel E. J. Geesen and Mick Geesen met the plane in Denver carrying the first Limousin semen into the United States. They inseminated a black baldy cow that same afternoon. Mark Curry, Geesen Ranch foreman and AI technician, is at right.

of Hereford, Angus and Charolais cows and cross-breds that gave a random selection of females for the Limousin test. To the French breeders it was a real eye-opener to the potential in the American market.

Other cattle had already been bred to our bull in Canada, where they had direct access to the semen. I held one Canadian meeting on July 7, 1968, at the Palliser Hotel. It was an impromptu affair put together by Sherman Ewing and my old friend Frank Jacobs, editor of the *Canadian Cattleman* magazine.

I had known Frank since the early days of the Pan American Charolais Association, and his excellent article on the Limousin breed appeared in the October, 1968 issue of his publication and sparked major interest in that country.

Present at the initial Canadian meeting in July, 1968, were Tom Eggertson, Calgary; Tom Gilchrist, Milk River; Bill Hart and Sam Hector, of Prairie Breeders; and the L.K. Ranches delegation consisting of Charlie McKinnon, Neil McKinnon and Edwin McKinnon, whose combined cattle operation repre-

sented one of the largest Hereford herds in Alberta.

Also present were Dr. J. A. Newman, Livestock Section, of the Lacombe Research Station; George Ward, Arrowwood; Hans Uhlrich, Claresholm; and Frank Jacobs. I had some color slides of Prince Pompadour and explained we were just beginning to process semen through the Canadian government A.I. center at St. Hyacinthe, and I outlined the basis of the Founder Membership plan. From this small beginning and the early Founder memberships of Sherm Ewing and Neil McKinnon, the Limousin movement in Canada got under way immediately.

Doanes magazine in the U.S. ran a special article on Limousin in their October 1968 issue from a news release that I sent them. Other publications joined in and we began to get a flood of inquiries and phone calls.

Our biggest problem was the simple fact we did not have enough basic information about the cattle. Then, in response to our cry for help, Louis de Neuville arrived in November, 1968, to answer all our questions. Louis at that time was president and founder of ELPA and this was his first visit to the United States and Canada.

We called a special International Conference to take advantage of his visit, and a Board of Directors meeting to work out the details of the by-laws and the general rules and regulations of the Foundation.

The Board meeting was scheduled for November 15, and the general conference for Saturday, November 16, at the Holiday Airport Inn, Denver. Announcements went out by mail, by telegram and by telephone. It would be our second meeting of the Board of Directors and the first time we had called a general membership meeting. We sent an open invitation to all interested stockmen who wished to attend.

There was a great response to our announcement and well over 100 interested people turned up—about three or four times our actual membership at the time.

As a special bonus, several of our early Founder members had just returned from France. They had visited the Limousin cattle country and we had their first-hand impressions for added commentary; H. A. McCoy, his wife, and Mr. and Mrs. Ben F. Price, Jr., had just returned from Europe and the Limousin country. They made very favorable comments on the breed, to supplement de Neuville's description of the cattle.

Then, in the middle of the discussion, Dick and Margaret Shaw walked in fresh from the Limousin country also. Dick and his wife had both taken Founder memberships and were enthusiastic about the Limousin possibilities in the U.S. As a long-time Hereford breeder, Dick Shaw had seen a number of things he liked about the golden cattle of France. His comments were particularly interesting to the many Hereford people now involved in our project.

Adrien de Moustier and his cousin, Antoine de Moustier, were also there to complete the roster of illustrious Frenchmen. The discussions with Louis de Neuville not only added greatly to our knowledge about the cattle, but it built a real "bonfire" of enthusiasm among our burgeoning membership.

A good many more people arrived on Saturday, November 16 and we made them all welcome. At that time we had only about 30 paid Founder members with no active members yet enrolled.

It was truly an international affair with nearly a dozen visitors from Canada, three from Mexico and U.S. delegates and members from the Atlantic coast to the Pacific, from Oregon to the Bahama Islands. Several who came simply to look and listen handed over a $2,500 check before they left.

It was also the beginning of a close association with Louis de Neuville that has continued through the intervening years, and it launched the fledgling Foundation into the middle of the international cattle breed industry.

At that Board meeting November 15, we had the first basic reports of our Performance Standards committee, headed by Sherman Ewing, and we also worked out the basic organization program for registration procedures in an open meeting in which several members participated.

Directors present at that time were: President Robert H. Purdy, presiding; Stephen Garst; H. A.

Louis de Neuville became the hub for information and encouragement as the Foundation and the breed grew in North America.

Sherman Ewing, Claresholm, Alberta, Canada was NALF Founder member #4. As chairman of the Technical Committee he was responsible for the writing of all the early directories on performance requirements for the breed in the United States. He was the first NALF vice president.

McCoy; Charles L. Moore; Ben Price, Jr.; W. W. Smutz, Jr., Bruce Waddle, Sherman Ewing and Jim Scott.

Founder members in attendance were: Burwell M. Bates, Robert Buckmaster, Leland Dudley, Stephen A. Lathrop, Adrien de Moustier, C. H. Thompson, Dr. J. P. Woodbridge and Charles McKinnon.

Also present as interested parties were: J. A. Garcia, Sr. and Jr., Brownsville, Texas; Wm. F. Hart, Sam Hector and Byron Palmer, all of Calgary, Alberta, Canada; Coy Hudson, Konawa, Oklahoma; R. D. Mohler, Maxwell, Iowa; Antoine de Moustier, Bov Import, Crecy, France; and our special guest of honor, Louis de Neuville, President of ELPA, and representative of the Herd Book Limousin, Limoges, France.

This was our second board meeting and it became a long discussion concerning the aims and standards of the new organization. Most of the officers and members had been directors in other cattle and performance associations and the talent and experience represented at this meeting helped to shape the basic proposals into a practical and constructive program.

On the following day, November 16, we held a general conference on the Limousin breed and its potential place in the American beef cattle industry. Louis de Neuville gave a detailed background talk on the cattle and their part in the beef industry of Europe. Dr. Robert de Baca, Iowa State University, outlined the American requirements for the success of any new breed arriving in this country.

Although not everyone registered for the event, the following group signed in: Mr. and Mrs. Roger D. Shaw, Thomasville, Missouri; R. D. Mohler, Maxwell, Iowa; Dr. and Mrs. James A. Scott, Fort Collins, Colorado; Joe F. Grose, Checotah, Oklahoma; J. Tipps "Tip" Hamilton, Kirley, South Dakota; J. A. Garcia, Sr., Brownsville, Texas; Robert Buckmaster, Waterloo, Iowa; J. J. "Bud" Prosser, Denver, Colorado; Leland Dudley, Hampton, Iowa; Mr. and Mrs. Ben Price, Jr., Reading, Kansas; Dr. R. M. Madsen, Omaha, Nebraska; Stephen A. Lathrop, Indianapolis, Indiana; C. H. Thompson, Hampton, Iowa; Richard M. Patterson, Denver, Colorado; George D. McLean, Toronto, Ontario, Canada; Mr. and Mrs. Don A. Pavel, Fort Collins, Colorado; Leo L. Chavez, La Jara, Colorado; Donald M. Culver, Denver, Colorado; Frank G. Watson, New York, New York; J. A. Garcia, Jr., Raymondville, Texas; Charles L. Moore, Coon Rapids, Iowa; Charles H. McKinnon, Calgary, Alberta, Canada; Pat Spurlock, Navajo, Arizona; Donald J. Kraus, Hays, Kansas; Byron Palmer, Calgary, Alberta, Canada; Sherman Ewing, Claresholm, Alberta, Canada; Ronald Wolff, Wheatland, Wyoming; Gene R. Brownell, Cortez, Colorado; Mr. and Mrs. Martin Allard, Loveland, Colorado; H. A. McCoy, Topeka, Kansas; Mr. and Mrs. D. N. Wright, Melba, Idaho; Thomas S. Christie, Harlowton, Montana; Dave Vecchio, Denver, Colorado; Mr. and Mrs. Burwell Bates, Konawa, Oklahoma; Loren R. Whittemore, Rush, Colorado; Mr. and Mrs. Coy Hudson, Konawa, Oklahoma; Adrien de Moustier, Paris, France; Antoine de Moustier, sur Crecy en Brie, France; Lyle Liggett, Denver, Colorado; Robert H. Purdy, Buffalo, Wyoming; F. Wallace Gage, Longmont, Colorado; W. F. Morey, Denver, Colorado; W. W. Smutz, Jr., Denver, Colorado; Dr. Robert de Baca, Ames, Iowa; Dr. and Mrs. J. P. Woodbridge, Pierson, Iowa; Mr. and Mrs. R. H. Smellage, Boulder, Colorado; Sylvia, Joanne and Star Smellage, Boulder, Colorado; Phil Van Dervoort, Wheatland, Wyoming; Don Steiger, Devils Tower, Wyoming; Forrest Bassford, Denver, Colorado; Sam Hector, Calgary, Alberta, Canada; William F. Hart, Calgary, Alberta, Canada; Louis de Neuville, Limoges, Haute-Vienne,

France; Bruce M. Waddle, Longmont, Colorado; Stephen Garst, Coon Rapids, Iowa; J. P. Smith, Amarillo, Texas; Richard Goff, Denver, Colorado.

Later, Forrest Bassford, Editor of *Western Livestock Journal,* asked me for more information on the Limousin breed, and in the January 1969 Stock Show Edition ran a feature story that gave us a tremendous boost in the cattle industry. This article on top of the previous press coverage we had received, aroused so much interest we were flooded with more than a thousand inquiries for that one month of January alone.

By this time, the idea of Limousin imports directly to the U.S. had begun to occupy the minds of a great many people. The fact no Canadian imported cattle could be moved to the U.S. for a period of five years seemed a bit severe, especially since a large part of the money involved in the Canadian program had been supplied by American breeders.

In several instances, the newly imported cattle had already been sold to Americans, which was specifically forbidden by Canadian regulations. No papers had changed hands, however, and the original importer maintained the animal on his ranch for a nominal annual fee until such time when the animal could be legally shipped to the U.S. In most of these deals there was also some agreed split on the calves, in the case of a cow, and if a bull, there was an arrangement on the semen which could be shipped to the United States.

In addition, there was already a bit of sub rosa bidding for permits by Americans that had reached substantial figures, ranging all the way from $3,000 (U.S.) to $12,000.

An increasing number of phone calls were coming in to my office wanting to know the story behind the rumors that were circulating by this time.

"How can I buy a Canadian permit and what will it cost me?" That was the question asked me almost daily. In order to explain the Canadian regulations and restrictions against resale to a foreigner, I put together a special bulletin on the import situation in a report to our Founder members, dated 21 February 1969. In a condensed form, this is what it said:

SPECIAL CONFIDENTIAL REPORT TO FOUNDER MEMBERS

"Every cattleman interested in Limousin cattle (or other so-called "exotic" breeds) wants to know how and when he can buy purebred French Limousin cattle and get them delivered to his own farm or ranch.

"For an American or a Mexican cattle breeder, the answer is pretty involved and the time and cost elements are very uncertain. The Canadian regulations are specifically designed to keep the imports there long enough to develop a basic breeding herd that will give Canadians enough seed stock to maintain a head start in Limousin. It's that simple and it goes back to the Charolais days when Canadian stockmen were on the tail-end of the buying line with that breed. The opening of the Canadian Import Station turned the deal completely around and they are now in the driver's seat.

"There are, however, some possible alternatives if you have the time and money to put into the project. To begin with, cattle have been imported to the U.S. from Ireland, Scotland and the British Isles for many years through an American import and quarantine station, set up and operated by the U.S. Department of Agriculture at Clifton, New Jersey.

These countries (plus Norway, Sweden and Japan) are considered free from Foot and Mouth Disease, Rinderpest, etc. and are not subject to the animal health restrictions against most European (and other) nations, which prohibit direct importation of any livestock.

This year, however, more Limousin will be imported into Canada, and it is hoped that as the supply of full French animals there increases, some modification of the Canadian regulations will occur.

This year, for example, there are 16 Limousin out of a total of 250 head now in quarantine that will be released in April (1969). These are divided as follows:

Ten head—six bulls and four females for the Canadian government for experimental work.

Three head—two bulls and one female came in for the Bov Import project at Rimouski. (We will handle the semen sales on these bulls on a similar contract to Prince Pompadour.

Two bulls came in for Prairie Breeders,

Calgary.

One bull came in to Dr. Phillips of Calgary, and has been leased to American Breeders Service.

These animals will be released from quarantine on Grosse Ile about the first week in April, then will be quarantined on their respective farms for an additional 90 days. At the end of that time (about the first week or two in July) semen will be available from these five privately owned bulls for shipment to the U.S. (We are told the Canadian government will not sell semen to private individuals from their bulls.)

What signs of encouragement are there for new avenues of entry? There is some interest in the USDA in the establishment of a U.S. Import Station, possibly in the Carribean on an American owned island, to operate in much the same manner as Grosse Ile does for the Canadians.

Another possibility is the establishment of a private corporation in some approved country where a breeding herd could be operated and the progeny imported through the Clifton, New Jersey, facilities. (This was later done by a group of Founder members in England and proved moderately successful.)

One or two other members brought Limousin into the U.S. by way of Sweden, in much the same manner. Mac Braly, a Charolais breeder from Ada, Oklahoma, had previously imported some 200 of the Charolais by means of a newly constructed quarantine station on the French owned island of St. Pierre, just off the coast of Nova Scotia. This island and an adjoining one were the only remaining French possessions near the territorial waters of Canada or the United States.

This unit was soon incorporated into the quarantine program of the Canadian government in order to expand their import facilities as the demand for European imports increased. The same restrictions then applied to cattle entering Canada through this facility.

The report went on to explain the availability of selected breeding stock in France and the restrictions imposed by the French health regulations which required all young calves be vaccinated for Foot and Mouth disease before weaning. This made it neces-sary for potential export calves to remain unvaccinated until they were ready to enter the French export quarantine facility at the port of Brest for their initial health tests.

We outlined for the members several approaches to the import project that had been discussed by some of our members and offered them any and all information that was available to us. We were also able to quote going prices being asked by new import holders in Canada for "partnership" interests. Some of these deals enabled the American partners to select the animal in France, pay the French directly for the animal, then later settle for the import expense with the import license holder.

As far as the Canadian government was concerned, however, the imported animal was still owned by the original permittee. It was a complex situation, but I received numerous queries from Canadians fortunate enough to draw a permit. Some were cattlemen and some were not, but as long as they had a permit they were in the cattle business, and they were looking for partners with money.

We passed on many names to our U.S. members in the early years of the import program and let our members make their own arrangements concerning cost and subsequent handling of the bulls or heifers involved. It was soon established that any progeny of the imported animals could be sold or shipped to the United States. This took some of the heat off the future import outlook. It also took some of the initial gamble out of the "partnership" propositions. And best of all, it materially improved the financial standing of some of the smaller Canadian cattle outfits.

In the newsletter I pointed out any program that would help bring additional Limousin into either Canada or the United States would help the Limousin Foundation in the long run and we would provide all the import information we had available at any time. The Foundation, however, would not be involved with any import project, but any member or group of members could act on their own as individuals or partnerships. We could give them the going prices, for example, that breeders in France were asking for calves, and refer them to Louis de Neuville or COFRANIMEX, the French export office for further help and advice.

By March 1969 the Foundation had established a working system of committees that included some of the most experienced cattle breeders in the country, plus an impressive array of some of the leading

Steve and Mary Garst, Coon Rapids, Iowa represented the extensive farming, seed and cattle business of Garst Farms. Steve served on the NALF board and was Founder Member #7. Garst Farms was #13 on the Founder list.

animal scientists from Canada, the United States and Mexico as technical advisors. These committees and their advisors had spent much time and effort to outline the basic tenets of a performance oriented program to promote and guide the development of the Limousin breed in America.

By mid-March we were ready to submit these final suggestions to our members for discussion and approval. On March 21, 1969, I sent a night letter by Western Union to every member announcing a special membership conference for April 18 and 19 to review the various committee reports and to hear suggestions, ideas and comments from a distinguished group of speakers.

Once again, Louis de Neuville, our dedicated Limousin consultant, flew back to Denver. Adrien de Moustier was here, and many distinguished experts from other countries. Among them was Dr. Jorge de Alba of Mexico and South America, who led some of the first Limousin off the ship that brought them to Uruguay a decade before our meeting. By a coincidence, Dr. Robert de Baca, who was chairman of our Technical Advisory Committee, had been in Uruguay at the time and was a witness to this event. Bob was also at our meeting as an advisor on our data processing and recording program.

Dr. Roy Berg of Canada, who was in charge of the test program for the 10 Limousin animals imported that year by the Canadian government, was

there to speak on that project. He and Sherm Ewing gave a complete report on the Canadian import outlook as it affected Limousin cattle. Bob Purdy and I made a report on our recent trip to Washington, D.C. to discuss the possibility of improving or expanding the U.S. import quarantine facilities for direct imports at Plum Island. At that time the outlook seemed rather dim.

Bill Smutz made a report on the proposed electronic data processing program for handling the Limousin registration, recording and performance program as our board and advisory groups finalized the details, with Bob de Baca commenting on the various approaches to the over-all program.

Our keynote speaker that day was Forrest Bassford, a man who had watched the development and progress of the Foundation from its very inception. His editorial support had been a big factor in the rapid growth and acceptance of our basic idea.

Another speaker on the program was Cecil Hellbusch, Agricultural Relations Director of Safeway Stores, and a widely recognized expert on volume retailing of beef.

The balance of the day and the following morning were occupied with discussions of proposed requirements for both progeny registration and performance data information. It had been established that the Foundation would make individual animal performance data a part of every certificate of record, whether crossbred or purebred. The debate arose about the minimum requirements in weaning or yearling weight, plus such factors as calving problem ratings, deaths and other critical data in the herd reproduction cycle.

One group maintained that weaning weights and daily rate of gain requirements should be high, while others pointed out the wide variations in weather, range or pasture conditions prior to calving and similar factors throughout the United States and Canada, made it difficult to make comparisons by numbers alone. The primary object on the 50 percent first cross calves was to get them recorded and work from there for herd improvement.

It was agreed no 50 percent bulls be recorded, but as much performance data as possible be gathered on these animals. It was also urged that steer carcass and gain data be gathered for future dam and sire ratings.

Later it was agreed 50 percent bulls should be recorded for two reasons: 1-It would help to gather performance data on these animals; 2-Their use as pick-up bulls on halfblood heifers would help to maintain the percentage of Limousin in those females that were missed by A.I. breeding. It was also some help as a marketing device for selling halfblood bulls in the early stage of the program to identify the original sire used if the bull produced unusually good calves.

All of these elements required a lot of discussion to determine the best program for the most people. One of the principal points of difference was in the use of the term ''purebred.'' At what point did a bred-up animal become the equivalent of a purebred. For years the Charolais association had decreed that five top crosses were necessary to advance to the purebred status. Our geneticists, however, pointed out that the fifth cross added only about three per-

cent additional Limousin blood to the animal and that quality and performance were far more important than this fractional increase.

As a result of this discussion, it was recommended the time-honored system of terming each advanced cross, such as one-half, three-quarter, seven-eighths, etc. up to the 31/32 degree that the Charolais used, be dropped in favor of the decimal system and show the various crosses as 50 percent, 75 percent, 87 percent, 94 percent and 97 percent. This was adopted by the membership at this conference.

Another difference in opinion developed over the use of the term ''purebred.'' The Canadians, quite naturally, who owned full French imports that were true purebred in every sense of the word, resented the competition of these home-grown purebreds, even though they were some years in the future.

Furthermore, the geneticists pointed out animals produced through four or five top crosses in a rigid performance selection program could very possibly be superior to an imported animal, despite his century-old record of bloodlines.

In the years to come this problem worked itself out by the adoption of the term ''Full-French'' to distinguish one type of purebred from another, a simple matter of cow country terms that proved to work quite well. And, from a standpoint of percentage value on a registration certificate, these were the only animals that could honestly show a 100 percent French bloodline. In time too, the elements of performance background did assume a greater influence, as our scientists had predicted.

Dandy NIM 3, by Noel and bred by Chastanet was used heavily in North America. His get contributed much to the breed history winning steer and performance shows in inter-breed competition.

Diplomate NIM 2, by Alaska and bred by Bourbon, was one of the thirteen "D" bulls to come to North America. Both Diplomate and Dandy were imported by Bov Import.

Chapter 3
Growth and Development

"The First Annual Meeting of the North American Limousin Foundation, held on Thursday, July 24, 1969, at the Brown Palace Hotel, Denver, Colorado, was opened officially at 10:15 AM by Executive Vice President Richard Goff, who welcomed all members present, especially those from Canada, as well as Adrien de Moustier, of France, and Dr. Jorge de Alba, of Mexico. Each member was then asked to stand and introduce himself by name, town and state."

This was the first paragraph of the official minutes of the first "annual" meeting of the Limousin Foundation. To those of us who had worked so hard and long to get this operation under way, and to the group of Founder members, friends and observers in attendance that morning, it seemed an historic event. It meant we had been a going operation for one year.

Announcement was made of a special exhibit scheduled for that afternoon from 4:30 to 5:30, consisting of four halfblood Limousin calves from the Geesen Ranch. These calves were still on their mothers and only a few weeks old, but Mick Geesen was bringing them in from the ranch at Agate, Colorado for display in the parking lot back of the Cosmopolitan Hotel. Thus most of the members would get their first chance to see 50 percent calves by Prince Pompadour out of Hereford and Angus-cross cows.

The Geesens were making the 150 mile round trip into Denver as a special tribute to the other Founder members of the Foundation, many of whom had calves on the way but had not yet had much chance to see the results of our first semen shipments the year before. It was a popular display and everyone expressed thanks to the Geesens for their efforts.

President Robert H. Purdy presented a general report of the Foundation's activities in its first year of operation. He pointed out that from a nucleus of 15 members who first met a little more than a year ago, we now had 80 Founder members, plus 13 active members and 9 junior members, two new categories that had just been announced.

He said promotion had by necessity been limited because it was tied to semen production, but our publicity had produced a tremendous flood of inquiries. He reported in the first month of the year, January 1969, we had received more than 1,000 letters requesting information.

Although semen shipments from Canada had been delayed for several months by unexpected U.S. Department of Agriculture restrictions, we had been able to ship 4,227 ampules of Pompadour semen by the end of our first fiscal year, May 31, 1969. By June 30, a total of 8,349 ampules had been shipped; almost as much in June as in our first year.

He explained actual production from Prince Pompadour had been 9,561 ampules for the first year, very close to our initial estimate of 10,000 ampules. Moreover, six weeks' production had been lost due to handling problems at the government processing plant in Canada. In addition, an unforeseen ruling by the U.S. government had delayed shipments into the U.S. by 90 days.

Now, he added, Adrien de Moustier or Bov Import, Inc. his Canadian company, had received two more Limousin bulls this year from France, Dandy and Diplomate, and the Foundation would be receiving semen from these bulls also. It was because of the previous semen shortage that the Foundation had not initiated the $100 Active memberships until this time. Founder members who paid their initial $2,500 fee for a priority position on semen had been rationed until recently. He pointed out that Steve Garst, Dick Shaw and H.A. McCoy had each taken two Founder memberships in order to get a double allotment of semen.

Purdy reported the Foundation was expanding rapidly. Receipts up to June 30 amounted to $241,464, with disbursements at $191,515, leaving a cash balance of $49,939. He cautioned that although this looked like a healthy position, we had major expenses involved in setting up a comprehensive system for handling the performance and progeny data our program would require. In addition we were starting a moderate promotion plan to sell the cowman on the value of using Limousin bulls and F-1 heifers.

He spoke of the future for the Limousin breed, pointing out that more efficient cattle and more efficient management was a necessity in the cattle industry. Cattle prices at that time were no higher than they had been 18 years before, yet during those years wages had increased 94 percent, consumer medical costs had gone up 98 percent, while cattle prices to the beef producer had remained constant.

He said the percentage of after-tax consumer dollars spent for beef had declined from 26 percent in 1946, to 20 percent in 1960, and to 17 percent in 1969, the lowest level in history. He stressed the importance of crossbreeding more productive, hybrid cattle in the future and said the F-1 cow was now the most important animal in the beef business today. This Foundation, he declared, offered cattle producers a head start in crossbreeding. We had an outstanding breed from a genetic standpoint that would fit the needs of the cattle breeder and the beef industry.

He said this breed and this organization was tied to performance from the sperm to the consumer, and that no other breed organization had ever had that opportunity before. Also, he said we had an outstanding promotion theme this year with the Limousin—"a genetic short cut to a better beef carcass."

Bob then asked me to outline some of our promotion and marketing aims for the coming year. I told them we estimated more than 6,000 cows had been bred to Prince Pompadour at this date and we expected another 10,000 to 15,000 to be in calf to him by next spring. With the addition of the two new Bov Import bulls, our supply of semen for sale by the Foundation would be greatly increased.

In addition, Bill Hart, of Prairie Breeders, Calgary, Alberta, had two new bulls just released from quarantine and American Breeders Service had one Limousin bull that would be in production shortly. On the basis of past performance, therefore, Limousin breeders could look for at least 50,000 ampules or more of semen available commercially for the coming year.

However, it appeared that Limousin semen might be in short supply for at least a year, and possibly two years, because of the widespread interest in the breed all over the United States and Canada. The critical point would be the number of Limousin bulls imported to Canada during the next year.

Our promotion for that first year was almost entirely free publicity based on news releases and articles I had written for a number of publications. The first article appeared in *Canadian Cattleman,* after editor Frank Jacobs and I put some material together.

I wrote another feature article for Forrest Bassford of *Western Livestock Journal* on the breed and the organization, which he ran in the January 1969 issue. These articles were a major factor in arousing interest in the United States and Canada. Both publications carried a great deal of prestige and authority in their respective countries, among the commercial cattlemen as well as the purebred breeders. The resulting interest began to show up at once in the inquiries for information we began to receive.

Other publications carried news stories on the new Limousin organization and its photogenic breed of cattle. Editors of some of the top publications in the agricultural industry were planning feature stories, among them *Successful Farming* and *Farm Journal. Doanes* magazine had just run a story in their monthly magazine on farm management, and *Farm Quarterly* had an article in production.

In summary, I mentioned my wife, Jane, and I had made a trip to the French Limousin country to see the cattle and gather data on the history of the breed, the French method of raising and marketing Limousin, and the handling of carcass beef in Europe. I had taken more than 500 color slides for use in our promotion work and at meetings around the country. I reported we had visited M. Louis de Neuville and many of the other leading breeders, particularly those in ELPA.

As another phase of promotion for the breed we donated 150 ampules of Pompadour semen to the new Meat Animal Research Center near Hastings, Nebraska, which was just starting their first crossbreeding program using semen from Continental breeds on Hereford and Angus cows. We heard more on this the next year.

Wally Gage and Bob Purdy had also made trips to Europe and England to look into the Limousin cattle and see what the possibilities were for setting up a possible alternate route for imports through some of the countries previously used in the importation of Charolais from France to the U.S.

Wally gave a general outline of some of the inquiries being made to circumvent the restricted approach through Canada. He particularly wanted to point out the Foundation itself would not be involved in any import project, but would encourage and help those members already working on this situation.

He mentioned Founder member Bob Smellage was working on a project in the Bahamas; Bob Bowers was working with the Province of Manitoba, in Canada; one group was working with the Irish government; and still another had plans for a project in England. He talked with some of the Charolais people in France and they favored the passage through Ireland and England, or through Japan. He added one or two such programs were already in operation for Charolais shipments to the U.S. Purdy had firsthand information about the outlook for a proposed American quarantine station.

Bob said not much progress was actually being made on this program and information had already been covered in Dick's Foundation newsletter. He

Bob Brooks, Seneca, Missouri calved out the first Limousin three-quarter in the United States. Bob, pictured at left, is shown here with George and Fred DeMier, Miami, Oklahoma, who held the first Limousin sale in Oklahoma. Both firms contributed much to the founding and growth of the breed.

added the government cost estimates for the project kept going up and were now up to $4 million or more, about three times what it would cost a privately funded corporation to do the job.

The afternoon session opened with a panel of the members of the Technical Advisory Committee consisting of Dr. Jorge de Alba, Mexico; Dr. Robert de Baca, Iowa State University; Dr. Carroll Schoonover, University of Wyoming; Dr. Ray Woodward, American Breeders Service; as well as members Sherman Ewing, Canada and Bill Smutz, Denver.

Dr. de Baca said he agreed with Sherm Ewings' Performance Standards Committee recommendations that all American Limousin should be bred on a technical performance basis and performance standards should be maintained as each generation was recorded from 50 percent, 75 percent, 87 percent and then special standards for 94 percent and 97 percent (fifth cross) animals that are registered as purebreds.

He advised against the use of 50 percent bulls in general, but said if they were going to be used they should be tested and recorded. He pointed out the proposal to eliminate the word ''purebred'' and substitute the term ''registered'' indicated any animal in the fifth generation, or 97 percent, but that animals to quality for ''registry'' at 94 percent

Prince Pompadour pictured in Las Vegas, Nevada at the 1969 International Cattlemen's Expo. Many cattlemen went home and joined the Foundation following the exhibit.

(fourth cross) had to meet high performance standards which would be set by the Directors. He added it would be necessary to agree on these points before the computer programming was completed. He also urged the organization to make every effort to get Limousin-cross steers on test from approved sires, preferably in a central location and under controlled conditions and feed rations.

Dr. Schoonover agreed with de Baca's statements and added he urged the Foundation to get into carcass work as soon as possible. He felt carcass data by sire groups was vital to those interested in the breed. He mentioned the Beef Improvement Federation (BIF) had recommended relating growth rate to carcass value in terms of total retail cuts per day of age.

At that time neither he nor I knew this prophetic statement would produce a Limousin carcass within the coming year that would set a new American record for cutability in a beef animal.

Dr. de Alba was introduced as the man who had brought the first Limousin to South America, and he commented he had been a Limousin enthusiast since 1960. However, he cautioned against accepting the Limousin as the ideal beef breed and felt we had to continue to improve the breed by selection to fit our specific market and production requirements. He also endorsed the program of high performance standards for registration. He added our carcass grading requirements for high marbling

was a major cost factor in American beef production, and he doubted it was as desirable as we believed.

Bill Smutz made a slide presentation of the various types of input and report forms that would be needed to collect and analyze the combined progeny and performance data being discussed for the Foundation standards. He explained the numerous aspects of data processing with the latest equipment then available and how the information might be programmed and handled as the organization continued to grow.

It was explained that standards and requirements for performance qualification, recording and registration had to be established by the Board before the final data program could be completed. In answer to further questions, it was explained changes could be made in the program if performance limits were changed in the future, but the basic approach had to be established and the data base built around these regulations.

Oftentimes the discussions became very involved, both from a technical standpoint and from the idea of what was practical and acceptable for the breeder. Our objectives seemed clear enough during preliminary discussions, but computer science, artificial insemination controls and the idea of an international breeding program were all new to most of our members. Even those who had been involved in

Charolais cattle breeding had not faced the requirements of performance data as a prerequisite to recording or registration.

At a later time we were invited to meet with the officers and data processing staff of the American Hereford Association to see if we could use some of the extra availability of their big 360 IBM computer. This unit annually handled the registration and transfers of millions of purebred Hereford cattle and still had ample capacity for additional thousands more. They would make us a very economical tie-in, from first appearances. Yet when we met with them and explained we needed codes for as many as 25 other breeds, and that any one animal might have a half-dozen or more breeds in the pedigree, as well as the need for birth weights, weaning weights, yearling weights and calving ease ratios, they began to see our problems. At that time their computer had a huge capacity for a relatively narrow range of data. We needed a much more limited numbers capacity, but a very wide range of detailed information.

Computer equipment today has the flexibility to handle all aspects of these data problems, but at that time it was difficult to explain and program the range of information we were discussing. No one had tried to combine all these elements before, as far as we knew. The dairy organizations included milk production figures, but they also had a much simpler program than ours, and few of us knew how to put our ideas together and make it understandable in those early policy meetings. As far as most of the Board members were concerned, we were talking up in the clouds. Furthermore, the "personal computer" type of equipment was then more than a decade away.

Iowa State University had been developing beef herd management programs for the big room-size computers then in use, and Bill Smutz had developed a system of ranch management programs for several cattle operations in the United States and Canada.

Thus, Bill had the nearly impossible task of trying to explain the planning procedures of data programming to a group of computer illiterates, which we all were. The greatest problem arose in trying to estimate programming costs under such conditions. Money was always a major planning problem in such a complicated situation.

The Board wanted to see some estimates on the cost of setting up such a system and no one knew the cost because it had never been done in this way

Joe Hochhausen, Edmonton, Alberta, Canada (left) was the first president of the Canadian Limousin Association. Pictured right is Cecil Hellbusch, Safeway Stores executive, who arranged for cut-out data on early crossbred steers for the Foundation.

before. Finally, Bob de Baca suggested he arrange for one of Iowa State's programmers to come out and work with Bill Smutz for a month and see what they could come up with. This was approved and the Board went on to elect officers, re-electing Bob Purdy, president; Bob Bowers, vice-president; Charles L. Moore, secretary; and Bill Smutz, treasurer.

At this meeting it was decided I should devote full time to the Foundation. Until then I had been operating my advertising agency and handling the promotion and development phase of the Limousin Foundation as I did other accounts such as Franklin Laboratories and similar clients.

Bob Purdy and I discussed this possibility a week or two before, and I finally agreed. It meant I would have to take a fairly substantial decrease in income, but as Bob put it, "since I had started the damn thing, I owed it to the Founder members to get in and run it!"

The agency, Clair, Goff & Fremd, was a profitable operation, but my partner of almost 20 years, Don Clair, had just decided to retire. It was a time of transition, in any case. I would turn the agency operation over to my other partner, Ted Fremd, and get into the Limousin project full time. It was a major decision for me from a personal standpoint, however, for I had spent almost 20 years building that business from scratch.

The Board endorsed the proposal and we decided to consolidate most of the office operations into one office, so correspondence and promotion could be handled in one place. Previously Bill Smutz had been handling much of the office operation from his office downtown and I handled the promotion and field work from the agency.

Purdy had first recommended we consolidate the operation into one office. Later we decided to move into larger quarters in the Stockyards Exchange Building, where it is today.

Another major development that summer was the organization of the Canadian Limousin Association. Under Canadian regulations only one registry organization could be chartered in that country for each breed. Those who were already members of our Foundation felt they should form a Canadian association and the two groups could work closely together.

The Canadian group held its organization meeting on May 18, 1969 and made application for a charter under the Ministry of Agriculture. This was approved in June and that association was in business with Joe Hochhausen as its first president and Laura Palmer as secretary.

At the July Board meeting, Walt Shatto outlined the Canadian Association program and said they planned to coordinate their operation with the Foundation. If it was practical they would use our computer service, when it was worked out. He wondered about expanding our advertising into the Canadian area, and I reminded him the very first article about Limousin on this continent had appeared in *Canadian Cattleman* magazine, and said we planned to continue our connection there.

We discussed some additional ways of working together in the two groups and made a tentative agreement to have a joint Board meeting in Calgary at their inaugural meeting, when they set the date.

We were all interested in promoting the breed and agreed if they could get the French cattle into Canada, we could work together to expand and develop the market, which would be principally in the United States.

That fall the effort to promote a quarantine station for the United States had made some progress in Washington and a hearing before the House of Representatives Agriculture Committee was scheduled for mid-November in Washington, D.C.

Bob Purdy, Steve Garst and I went to Washington, and appeared before the Livestock and Grains Subcommittee to plead the case for a U.S. quarantine facility in behalf of American cattle producers on the basis of economic need for more efficient hybrid beef animals. In a written statement I outlined the benefits to the American beef industry of such a facility and that the Canadian monopoly on all continental seedstock made it difficult for the U.S. beef producer to compete.

The price of this purebred animal from France, I explained, cost a Canadian breeder about $4,500 to $6,500 delivered to his farm. He then offered to sell an interest in the animal to an American producer from four to five times this amount. Furthermore, Canadian restrictions on these imported cattle meant they could not be exported to the United States for three to five years.

In the event I was challenged on such a statement, I had with me several of the many letters we had received from Canadian owners of imported animals offering fractional interests for sale. The most recent one from a lucky winner of a permit in the previous year's import lottery offered a one-eighth interest in his bull for a mere $9,000. He added he had two more bulls in quarantine now and would be offering them for sale upon their release from quarantine. For the $9,000 a buyer would expect to receive 12.5 percent of the semen the bull produced and share the cost of maintaining the animal and processing the semen.

The bull had been sent directly to a Canadian A.I. center and there was no indication the owner had any interest in cattle breeding or the Limousin breed. He simply wanted to cash in on his lucky draw.

This sort of "marketing" plan on the Canadian cattle had a noticeable effect on the type of people who came into the Limousin project in those early days. I talked to any number of outstanding Hereford, Angus and Shorthorn breeders who were genuinely interested in Limousin cattle but who refused to play the game of roulette with the Canadians. As a result the initial phase of our activities ended up in attracting a high percentage of wealthy breeders who were in the cattle business as a tax benefit sideline.

That in itself was not in any sense a handicap, for the Angus and Hereford breeds had both attracted the big money people from their earliest days. Cer-

Dr. Robert de Baca addresses the 1970 organizational meeting of the North American Limousin Foundation. At the speakers table are Joe Hochhausen, Jim Baldridge, de Baca, Dick Goff, Louis de Neuville, Dr. Carroll Schoonover and Dr. Lavon Sumption. President Bob Purdy is hidden by the speakers rostrum.

tainly, many of these people made a great contribution to our breed and our organization, when they were knowledgeable about the industry.

Our explanation to Congress seemed to produce eventual results, however, for a bill was passed to establish a U.S. livestock quarantine import station in February, 1970, and officials of the USDA said they had begun a search for a suitable location. Fundng for the project was limited to a location and planning study only. No one would hazard a guess on an approximate date for actual importation of the first cattle. I observed in one of my newsletters that Canadians had planned, built their facility and received their first import cattle in less than a year. The USDA station planning and construction effort took nearly 10 years to accomplish the same job.

In December, 1969, we decided to promote Limousin at the first Beef Expo in Las Vegas, Nevada, a promotion set up by Jack Linkletter. They were going to put a lot of promotion money into this first show and wanted us to take part. It sounded like a worthwhile promotion and he assured me they had a world of money backing it.

I figured we might make a deal with Adrien de Moustier and the Canadian government to let us bring Prince Pompadour to the U.S. and this could be his American debut. This show was in December and, if we could keep him here for the next few months, he could be on display also at the National Western Stock Show in January. He would be the first purebred Limousin bull to set foot in the United States.

It turned out to be a fairly expensive proposition for us. The Canadian government wanted an iron-bound contract that the bull would be returned at the specified time, plus a $100,000 performance bond. I told them we would agree to the terms, but I wanted a six-month time period for his visit. I pointed out he would be down here in much warmer weather than his Quebec stable enjoyed, and I did not want to risk the need to return him during the bitter cold weather there in February and March. He was much too valuable an animal to expose to a sudden change in climate.

I thought if we were going to put all that money into this promotion, the National Western Stock Show would give us more exposure to commercial cattlemen than any other livestock event in North America.

It turned out to be an outstanding sales project for the breed and the Foundation. We arranged with International Beef Breeders to handle the job and to house the bull during his stay in Denver. Bud Prosser and Terry Carlstrom handled the hauling of the bull for the 2,700 miles from Rimouski, Quebec, to Las Vegas. They rested him several places enroute and gave him an extra several days at Denver, before taking him over the divide to Las Vegas.

The Las Vegas Expo was a success from a promotion standpoint. Several people who saw the bull at that event later went to Limousin and some of them became prominent breeders over the next few years.

We had the bull back in Denver and comfortably housed at International Beef Breeders for the National Western Stock Show in January, 1970. There he had a steady stream of visitors and I felt this exposure at the largest stock show in North America was well worth the time and money involved.

In addition, IBB had the bull producing semen all the time he was there and we built up our inventory by several thousand ampules during his visit. Since we did not have to pay import duty on this semen, it was an additional bonus to our promotion value. Certainly as a result of the promotion and exposure from this "tour," Prince Pompadour became one of the world's best known beef animals.

When he returned to his home in Rimouski that spring, we felt he had done an outstanding job for the Limousin breed and the North American Limousin Foundation. Louis de Neuville had attended the Las Vegas show and was able to convey an enthusiastic report on his appearance to the ELPA organization in France.

Adrien de Moustier came to the National Western and we went to visit "his bull." Both of us thought how much simpler things would be for our project if we could keep him right there for the rest of his life. But that $100,000 performance bond meant he had to return to his white barn on the south bank of the St. Lawrence River, and return he did in good health and top condition.

By the spring of 1970 we had quite a number of halfblood calves and yearlings in the pastures of our members. The bulls began to show up in test stations across the country, and reports on their progress began to reach my desk. One group in particular seemed to be doing a good job of daily gain performance at the BIAS bull test center in Sheridan, Wyoming. These were a pen of nine bull calves entered by Bob Purdy. They were out of some of his Charolais-Angus cross cows and all were sired by Prince Pompadour.

When the results were in on the 140-day test, his pen were the top performers in the test out of a total of 202 animals of various breeds and crossbreeds. One animal in particular stood out, his Number 909. This was a late calf, born June 19, 1969, but he was the best performing animal in the entire test.

He had gone into the test weighing 491 pounds and weighed out 140 days later at 1,014 pounds for a total gain on test of 523 pounds, an average of 3.75 pounds per day. His weight per day of age was 3.4 pounds, the highest of any other animal in the test of any breed or crossbreed.

Our second annual meeting was coming up in July and I needed a special event for that meeting. I had observed in several of the carcass contests I had been involved with in the past that the high gaining cattle had usually turned out a high yielding carcass. I wondered what the carcass on that bull would look like, especially since I had seen a number of bull carcasses being cut up in France. The French Limousin beef industry had changed from steers to bulls many years ago because of their higher yield and absence of fat.

I called Bob that afternoon and congratulated him on his pen's performance. I asked him what he was going to do with those bulls and he said he would probably sell them. I told him I wanted to buy the 909 bull for the Foundation and asked him to put a price on the animal. He stalled a little bit and said he had planned to use him in a breeding test, but he finally settled on $1,500. That was a high price at that time for a halfblood bull, but there had been none to sell previously from a performance test group that I knew of. I told him I wanted the bull and would send him a check that day. I asked him to leave the bull at the test station and I would come

The now famous "909 bull" pictured as he completed his performance test at the BIAS, Sheridan, Wyoming performance test station. He had an ADG of 3.75 and the highest WDA, 3.4 pounds, of any bull in the all-breeds test.

up and get him at the time he was a year old in June.

"What are you going to do with him, Dick?" he asked.

"Bob," I replied, "I'm going to butcher him."

"Butcher him!" Bob shouted back at me. "That's the best calf out of my entire '69 calf crop. You can't do that to him!"

"Bob," I said, "You and I have seen the French cut up purebred bulls in France that were a lot better than that one. I want to see how one of our best crossbred bull carcasses will compare with those French cattle. If we use anything less than your 909 bull we won't prove a thing. Same way with a steer."

I wanted to have that carcass on display at our annual meeting in July. I explained I had taken a measuring grid with me on my last trip to France and had seen bull carcasses there with ribeyes that measured over 20 square inches.

I knew Bob's cattle from our Charolais days, 10 years or more before we got into Limousin. This looked like the opportunity of a lifetime, and I hoped we would really find out something. I told him Dr. Schoonover at the University of Wyoming had been doing cut-out tests on Hereford-Angus cross bulls for the past five years or more and that he was especially interested in this type of beef. I planned to take the bull down there and ask him to process the animal and do a complete analysis for us.

I reminded him we had set this group up as a research organization and I wanted his bull to be our first research project. Maybe we could show our members a kind of beef they had never seen before.

"Well, okay," he said when I finished, "And good luck!"

I told him I would stop at the ranch on my way to get the bull in my own pickup and would haul him from Sheridan to Laramie myself. I wanted to see Bob's heifer mates to those bulls, too, and would get some photos of those cattle for promotion purposes at the same time.

It was about the middle of June when I finally got to the BIAS Test Center at Sheridan and saw the young bull I had just bought. He was an impressive animal to sacrifice on the altar of Limousin research. I shot some color and black and white photos of the bull and hoped they would insure his fame for posterity. He was a very heavily muscled animal, dark red in color and a good disposition. Just the kind of critter that would make an outstanding breeding bull. I looked at him and hoped I knew what I was doing. By this time he weighed 1,200 pounds and he really filled up the stock rack on my three-quarter ton Ford pickup. I waved goodbye to

When "909" was slaughtered at 363 days of age he had a 21.2 square inch rib-eye. Louis Swift and Jim Baldridge view the carcass that inspired the breed theme—Limousin the Carcass Breed. This was the first Limousin carcass cutability demonstration in North America and was conducted by Dr. Carroll Schoonover, University of Wyoming.

H.A. McCoy, Miami, Oklahoma was Founder member #11 and shared #12 with Ben and Ruth Price, Reading, Kansas. "The Real McCoy" is an enthusiastic booster and promoter of the breed and his favorites—Sooner and Hillary.

Bob who had driven with me from his 28 Ranch near Buffalo, and headed south for Laramie.

I had talked to Dr. Carroll Schoonover a week or so before and we arranged a date for our project. Schoonover's Meat Research Center at the University was beautifully equipped with everything needed to do the job. Furthermore, he probably had more experience with bull carcasses than anyone I knew of in the meat research field. His findings in the next few days might tell us what the Limousin cross would do.

At the University I unloaded the bull, found a motel and met Schoonover the next morning at the Lab. This was the moment of truth as the Spanish say in the bull ring. I had to turn away when Schoonover raised the rifle at that magnificent head in the killing chute.

When the carcass was up on the rail and the hide off, I began to feel better. That beef animal was something to see. Those hind quarters looked exactly like the carcasses I had photographed at an abbatoir in France the year before.

Later when Schoonover took his big knife to rib down the right side of the carcass, I took one look at that ribeye and knew this was a special event in the beef industry. Schoonover got his measuring grid and made at least five or six counts on that cross-section. I could see he found it hard to believe that a yearling bull, exactly 363 days old, could have produced a loin muscle like that. Finally he announced it was 21.2 square inches.

He recommended we keep the two sides in the cooler for at least a week in order to let the beef set up—the usual procedure in their cut-out tests. I said I would be back to watch that part of the project and to shoot pictures of the various primal and retail cuts as the work proceeded. In the meantime I went back to Denver, called Purdy that evening and gave him a preliminary report on his bull carcass.

It was about 10 days later when Schoonover planned to complete the cut-up of the carcass. In the meantime I decided I should get a couple of additional members of the Foundation in on the event, so I called Jim Baldridge in North Platte and Louis Swift in Fort Collins and asked them to meet me in Laramie to see this unusual carcass. I told them I thought it would be a real education in Limousin carcass potential.

Jim had a wide background in carcass work with the Angus Association and the Berkshire Hog Association during the development work on the "meat-type" hogs. Louis Swift came from a family that had dominated the meat-packing industry for a century or more. I couldn't think of two more knowledgeable people to observe the cut-out results with me. Besides, Jim was to be one of our principal speakers at our up-coming annual meeting in July.

When the day came for the cut-out, both Jim and Louis were there. Jim had brought Dale Runnion and Dean Jacobs, both meats majors in college, along with him. No one had ever seen a 21-inch ribeye on an American-bred beef animal before. As the cut-out proceeded the amount of beef that came off of that first side was, as one of the young graduate assistants expressed it, "awesome."

As each primal was broken down the pile of red meat stacked up on the scales was astonishing. Schoonover kept looking at the figures and double checking his results. He didn't say much. Finally, when the figures were all totalled up, I ventured to ask him how this carcass compared with some of the other bull carcasses he had worked on.

"Well, Dick," he remarked quietly, "There is no comparison to this animal and any other we've ever cut out." I had the feeling he felt this animal was a freak and we might never produce another one like it. I mentioned I too was surprised at the results, although I pointed out I had seen some carcasses in France that gave almost identical figures.

The astonishing thing to me was the fact our first halfblood yearling could equal the cut-out of the good commercial Limousin bulls in France. If this was an indication of what we could expect in the future, then the Limousin could make a major impression on the beef trade in years to come.

My news release on the results of this first NALF research effort—the "909 Project"—was widely used in the agricultural press and attracted widespread interest in the breed. I had a lot of phone calls to ask if this was true and whether others could expect those results. I explained this was an exceptional animal for a halfblood, but the Limousin in France were producing these results consistently. We could expect a wide variation in carcasses depending on the breeding of the dam, the feeding program and the many other factors. Correlation between high gain performance and high carcass yield was well known, but Limousin had added a new dimension to that factor and we had a lot to learn about the breed. This was just the first experiment we tried and they could expect more information in the years to come. It was a major attraction for our July annual meeting.

At that time we had about 85 Founder Members and about 25 regular or Active Members, a new category we had just opened up as the semen supply improved. The "D" animals had come out of quarantine in Canada the previous spring, and now the "E" bulls were just released. (The French used a letter system to indicate the year each animal was born.)

The Limousin that had come out of quarantine the preceding year included six bulls and four females consigned to the Canadian Department of Agriculture for research purposes only. These cattle were at the Brandon, Manitoba, research center and semen from the bulls was not available to Canadian breeders.

A bull named Decor, imported by Walter Phillips, Calgary, Alberta, was leased to American Breeders Service A.I. Center near Calgary.

Two other bulls, Danseur and Dary, were imported by Sam Hector for Prairie Breeders A.I. Center, Calgary.

Decor (x Beta out of Alcove x Ubu) was imported by Dr. Walter Phillips, Calgary, Alberta. He was bred by Pompadour. From available records more Decor semen was sold in the early years than any other Limousin bull.

Two bulls, Dandy and Diplomate, plus one female, Danseuse, were imported for Bov Import, Inc. by Adrien de Moustier, and these were at Rimouski, Quebec. We had previously contracted for semen from these two bulls to be sold through the Limousin Foundation. These five bulls and Prince Pompadour had been supplying the entire North American continent up until that time. Now the picture was changing fast.

By the summer of 1970 another 30 bulls and some 43 females were released from quarantine in Canada, and the Limousin build-up of purebred French animals was under way. The previous year there had been over 2,000 applications for import permits by Canadian breeders, but less than 300 had been issued. Out of that group some 75 had chosen Limousin. It was predicted by some Canadians the percentage of Limousin would more than double for the permits issued that year for release in 1971. In addition the Canadian government was more than doubling the capacity of the quarantine facilities by adding another station on the French island of St. Pierre, just off the coast of Newfoundland.

At our July 1970 annual meeting (our second) attendance jumped to a record high. Prior to that time we had about 87 Founder members and 25 regular members. More than 300 people came to the "Second International Limousin Cattle Conference" at the Cosmopolitan Hotel in Denver. Our new Active membership category increased by a hundred or more.

Bob Purdy, president of the Foundation, opened the conference with the statement Limousin had become the most talked about breed in America. He said the Limousin was a distinct breed of its own with production and efficiency factors such as higher weaning weights and daily gain capability as well as high fertility, easy calving ability, extra feed efficiency and a truly exceptional carcass quality and cut-out value.

"These are all qualities the Limousin has already demonstrated to a pronounced degree, and they are an essential part of the overall production capacity of an entire beef cow herd," he declared.

He urged cattle producers to consider the total production capacity of their herd as an entire unit in evaluating beef production qualities and gross income potential. He warned against getting "hung up" on one feature such as weaning weights and ignoring some of the other equally important elements such as fertility, calving percentage and ease of calving.

Purdy said his own crossbred Limousin heifer calves weaned at 658 pounds, while his Charolais-cross heifers averaged 615 pounds. The comparative

Jim Easland, DeSmet, South Dakota breeder, was the country's greatest membership salesman. He was Founder member #48 and went out and sold 48 memberships. He served on the NALF Board for two terms.

figures for his bulls were 672 pounds for Limousin crossbreds and 656 for Charolais crossbreds.

Nine of those Limousin-cross bull calves went into the BIAS Test Station at Sheridan, Wyoming, and produced the highest weight per day of age record out of the entire 202 head on feed, leading all other breeds and crossbreeds.

Purdy said the dressing percentage on the 909 bull was 68.35 percent with a ribeye area of 21.2 square inches, an apparent yield record for a yearling animal. He pointed out the USDA cutability index in that carcass was 58.35, which meant the carcass yield in trimmed retail cuts and lean trimmings was a remarkable 83 percent. This figured out to a carcass weight per day of age of 2.12 pounds. The ribeye area per 100 pounds of carcass weight was an astounding 2.75 square inches, he added.

He said the animal was a late calf, born June 19 and out of a three-quarter Charolais, one-quarter Angus. He pointed out this kind of production efficiency is an outstanding testimonial to the future value of the Limousin breed. "How much extra value the Limousin can add to our American cow herds we do not know at this time," he commented, "but this first example is most encouraging."

There is only one reason for any cattleman to crossbreed with Limousin, Purdy declared, and that is to make money. He said having calving problems and low fertility can easily reduce the net income of a cattle operation far more than low daily gains. He pointed out the features of the Limousin breed were all important to a greater net profit for the herd.

There was a noticeable feeling of growth and progress in the 1970 convention. We had an outstanding group of speakers and all were obviously impressed by the first year's calf crop, although very few had seen a purebred Limousin calf and not all had seen many halfblood calves.

Among those speakers who talked about Limousin and what it could mean to America was Louis de Neuville. He told about the work being done on ELPA's new bull test station, and the increased facilities in France for handling the export cattle to Canada and other countries. He also conducted a question and answer session on the Limousin breed from a breeder's standpoint that brought a lot of response from the audience.

Claude Morand-Fehr, Commercial Director of

COFRANIMEX, the French livestock export organization, and former chief of the French National Beef Marketing Syndicate, said from his observations in the U.S. the breed would prove to be ideal for the American cattle market. He added the breed was also attracting attention in other parts of the world and that exports last year were 153 head while the number this year was 450 head, with every indication that a comparable increase would occur in the coming year.

A panel on the problem of the Culard, or so-called "double-muscling" trait that had shown up in some beef breeds, concluded their discussion by observing that none of the Limousin cattle observed in this country or in France had shown any signs of the problem. On the panel were Dr. Tom Sutherland of Colorado State University, who had spent several years in France engaged in many phases of beef cattle research there; Dr. James Scott, Fort Collins, Colorado, veterinarian; and J.J. "Bud" Prosser, Mead, Colorado.

Jim Baldridge of North Platte, Nebraska, Founder member and breeder, urged a proven meat sire program for the Foundation and outlined what he called his "No Nonsense Nineties" program. Its principle features were:

A 90-day calving period for the entire herd, if one season per calving is used.

90 percent of the cows should calve each year.

90 percent of the cows should calve unassisted.

90 percent of the bull calves should weigh 600 pounds at 200 days.

90 percent of the heifer calves weigh 550 pounds at 200 days.

90 percent of the bulls should weigh 1,100 pounds at 1 year.

90 percent of the heifers should weigh 850 pounds at 1 year.

He had additional factors about grading and finishing at 1,100 pounds and carcass specifications to fit the current market requirements. He stressed the point that a sound breeding-up or crossbreeding program required severe culling procedures if the breed was to maintain and improve quality beef for the breed. He also advocated a continuing program of production and pedigree information that would be useful to breeding plans.

A panel of experts on beef production efficiency included Dr. Robert de Baca of Iowa State University; Dr. James Brinks of Colorado State University; Dr. Lavon Sumption, Lethbridge, Alberta; Canada; and Dr. Carroll Schoonover, University of Wyoming, the meats expert who cut up the carcass of the Purdy Limousin bull Number 909. He said of the animal, "You had a tremendous animal there—will you be able to produce more like him? I think you may, but you will have to put in more meat quality. Number 909 grew so fast—1,200 pounds in 363 days—that he had no chance to stop and put on finish. Add the marbling and the Limousin will be highly competitive in the beef industry."

Dr. de Baca pointed out the Number 1 trait for any calf is to be born—and live! A dead calf, he declared, is no help to beef herd efficiency.

Announcement was made of plans for the first Limousin sale to be held on opening day of the National Western Stock Show on Friday, January 15. I arranged for this event to be the beginning of an annual "Limousin Day" since that first weekend was traditionally the largest attendance part of the 10-day event. Jim Baldridge, National Livestock Brokers, was sale manager, James Scott, Bruce Waddle and Louis Swift were appointed to the sale committee.

Baldridge announced the sale was open to all members for consignment of 50 percent yearling heifers, both bred and open, and he hoped to locate a few halfblood cows with three-quarter calves at side. Later the sale was opened to a limited number of select halfblood bulls. Numbers would be limited to 200 head, and the entries must be recorded and have complete performance records, the committee announced.

Officers and directors for the coming year were also elected. They were: president, Robert H. Purdy, Buffalo, Wyoming; vice-president, Robert D. Bowers, Lake Francis, Manitoba; secretary, Burwell M. Bates, Konawa, Oklahoma; treasurer, John D. Moore, Newville, Pennsylvania; Richard Goff was named executive vice president; and James Scott and J.C. Easland were named director members of the Executive Committee.

Other directors elected were: James L. Baldridge, North Platte, Nebraska; Stephen Garst, Coon Rapids, Iowa; Walter Shatto, Calgary, Alberta; Sherman Ewing, Claresholm, Alberta; Charles L. Moore, Coon Rapids, Iowa; Bryant Harris, Marfa, Texas; Waldo E. Forbes, Jr., Sheridan, Wyoming; R. J. Steward, Keating, Oregon; and Joe Chase, Halliday, North Dakota.

At the business session, Ron Brown, our new systems director, announced plans for a new combined registry, recording and performance data program that would go into effect within the next three months. The four steps in the procedure were: 1—Herd enrollment; 2—Calving report; 3—Weaning report; 4—A post weaning list. The latter could be used for registration or recording of calves and it was predicted the new system would eliminate nearly 90 percent of the previous paper work involved in maintaining herd records.

It seemed a continuing problem to reconcile the complex handling of data for our members and the electronic programs and equipment then available. Before long we were in a time-share system that involved two major computers, one in California and another in Ohio. I never understood this part of our expanding operations, nor did most of our directors. We put our faith in the machines and the people who ran them. Terms like "down time" and "turn-around time" became recurrent headaches that could not be repaired with a pair of pliers and a screw driver. Programming was all done in a foreign language named Cobol, although so far as I knew there were no Cobollians available to come in and translate for us. They were not like the ancient Greeks or the long vanished Etruscans; they seemed to be part of some science fiction mythology.

I suggested one time we simply hire three or four alert young typists and rent some IBM typewriters in order to handle some of the back-log of applications and transfers on hand. But that was received as blasphemy. Once started on a computer program policy, it seemed there was no turning back. Time and money were irrelevant. Cows were being bred, calves were dropping, weaning weights were conscientiously recorded and the grim economics of the cattle business went on as it had for countless generations. This made it difficult to explain to an old cowman who needed to sell some heifers right away, that it would be a few weeks until his certificates came through.

In time, even computers can be whipped into shape and the new system began to work more efficiently. The problem during those early years was that applications kept increasing faster than capacity. Rule changes in procedure often caused unexpected problems, too. We had growing pains, and like a case of hiccups, everyone had a different remedy to suggest. Time and long hours of conscientious staff work, however, improved our production in this field.

Then a major public relations problem cropped up that threatened the Limousin reputation from a new angle. It went back to the 150 ampules of Prince Pompadour semen we had donated to the Meat Animal Research Center in Nebraska for their comparative research with other Continental and American cattle breeds.

These had been used to breed Hereford and Angus heifers and the resulting calves were now on the ground and the data had been compiled in a preliminary report that was released to a prominent livestock newspaper. The resulting story appeared on August 27 and reported Limousin had much more calving difficulty than either Simmental or Charolais calves. This was certainly contrary to most of our field reports, but it got widespread distribution immediately and my phone began to ring.

Since I had not seen a copy of the report, I called Keith Gregory, head of the test center and asked him to send me a copy of the data. Several other editors called me for comments and I stalled until I could see and study the information.

The Stockman publication, I found on reading the complete report, had simply skimmed the surface of the data. It was true Limousin had more calves rated difficult calvers on first-calf heifer cows than other crossbred calves in the test. There were several other vitally important factors that had been overlooked, however, and these were more important economically.

The report said 65.1 percent of the Limousin calves had required a calf-jack for assistance in delivery, versus 39.4 percent for Simmental and 60.5 for Charolais. The next column in the report listed calves delivered by Caesarean section. In this category the report showed 4.8 percent of the Limousin calves delivered by Caesarean, while the Simmental had 24.3 percent taken surgically. Charolais had 7 percent in that category.

A third factor was the number of calves born dead. In this column the report showed Charolais calves with 16 percent dead at birth, Simmental nine percent, South Devon ten percent and Limousin six percent. It was interesting to note too, I said in my report to the members, that 11 percent of the Angus calves were born dead.

In my newsletter I commented that a study of the data showed nearly all of the two-year-old heifers had difficulty calving at the station. In fact, out of the 413 calves born to these heifers 40 percent of all breeds and crossbreeds were pulled with a calf-jack. This applied only to first calf heifers and very little difficulty was encountered in the three to five-year-old cows.

Some time later I talked to a young man who was working there at the time this calving period occurred. I knew his family and knew he had grown up in the Hereford business and had seen lots of first calf heifers birthed before. I asked him what the problem was that first year. He told me most of the workers there at that time had very little experience in calving cows or heifers. Most of the older cows in the herd had calved first and because of shipping delays in getting semen, most of the heifers were bred late the previous year. This was especially true of the Limousin semen. As a result, the attendants became very impatient with those young heifers and many of them were pulled when a little more time and patience would have given them a chance to calve by themselves. He grinned and said this was especially true between midnight and morning, when everybody was tired and didn't want to sit around in a cold calving barn any longer.

I also pointed out in my newsletter that one of our members had called to say he had calved out 158 Red Angus heifers that had been bred to Prince Pompadour that same spring and he had pulled only two of them.

I said in the report also, in all fairness to the MARC operation, this problem was probably due to the lack of experience with these kinds of cattle and the overall "newness" of the research operation. Certainly their later experience with Limousin and the other Continental breeds showed marked improvement in their management operations, and they have produced some outstanding work in the subsequent years.

We knew growing pains and new techniques could cause problems too, so we put that incident down to learning and experience. Fortunately, my special bulletin to our members about the incident seemed to answer nearly all questions and we went on to other things.

That fall we promoted a tour of American breeders to France to see the Limousin foundation herds on their home turf and it was a great eye-opener to those who made the trip. Those herds of "original" Limousin look different in the hills of the Limousin country than they do anywhere else in the world.

By the first of the year, January 1971, we were well established in our new quarters on the third floor of the Livestock Exchange Building and preparing for our first display and sale at the National Western Stock Show.

Jim Baldridge and his committee were assembling a good selection of heifers. The group also considered a few outstanding crossbred bulls.

As a result of our previous tour to France, Louis

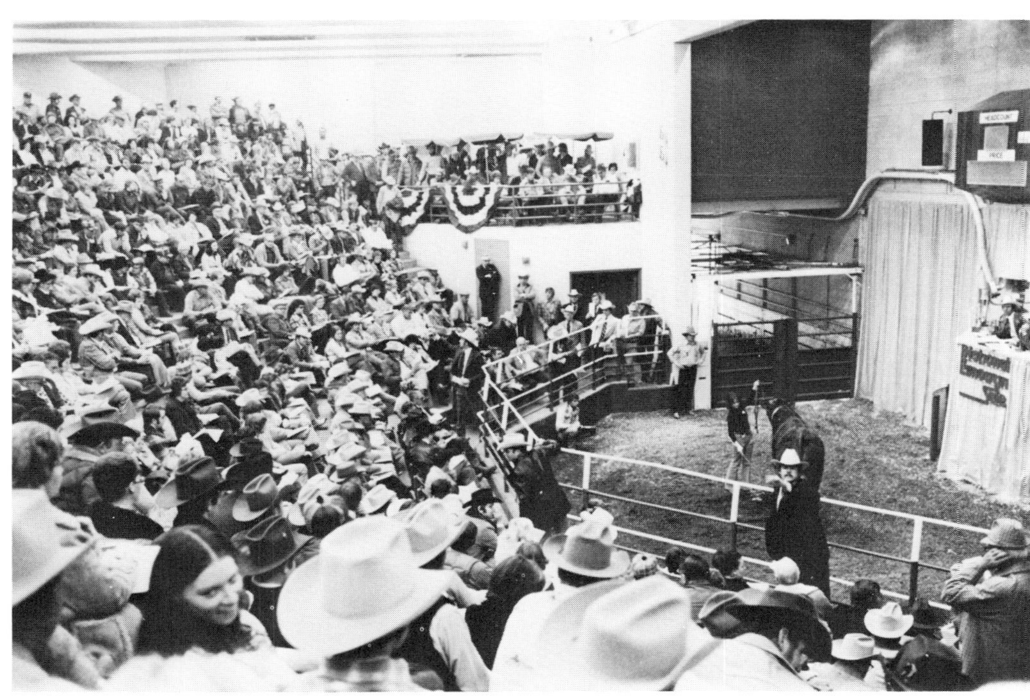

The first National Sale held in the spacious National Western sales facility in January 1971 drew a record crowd. Buyers came from 12 states and two Canadian provinces.

A halfblood female, bred by Dudley-Thompson, Hampton, Iowa, was the top selling female at the first National Limousin Sale. The event was held during the 1971 National Western Stock Show in Denver, Colorado. Shown with the $10,100 female is Leland Dudley at the halter and visiting French breeders Regis Coudert, Emile Chastanet, Yves Lemaigre-Dubreuil, Pierre Pechdo, Claude Mercy, Paul de Monvallier, Olivier Vandermarcq, Olivier de Rochebouet, Louis de Neuville. John Frezieres, representing the buyers group, LFB Corporation, is at right.

de Neuville arranged for a group of French Limousin breeders who were planning to attend our first sale of crossbred Limousin and to see the National Western Stock Show, one of the largest events of this kind in the world.

Some of the most prominent French breeders, most of them members of ELPA, arrived with Louis for our Stock Show and Sale. They were fascinated by the crowds, the large numbers of American cattle, and the time and effort that went into fitting the animals for show.

They were especially critical of the finish on all the show animals. "Too fat!" they all said. Their greatest astonishment came at the opening of the Limousin sale, however. Louis de Neuville was the only member of the group that had ever attended an American livestock auction before. When Jim Baldridge opened the sale with his auctioneer's chant and a high decibel level on the sound system, they looked at each other in open-mouthed astonishment. To them it sounded like the war chant of a tribe of wild Indians, one of them remarked later.

When the first heifer, consigned by Dudley-Thompson, was sold for $10,100, and they had figured out the price in French francs, they were impressed. I was watching the group from the sideline and I could see the wheels turning in their minds. I figured the price of purebred cattle in France was sure to go up on their return.

The next heifer brought $7,600 and then the price stabilized for a time between $2,000 and $3,000. The first 13 bred heifers averaged $3,223. By the end of the sale, 90 open heifers averaged $1,178, and the total of 103 lots was $148,075.00 for an overall average of $1,437.62.

It was a major success and the big sale ring in the yards was packed all through the afternoon. The French breeders were impressed with the technique of selling and the volume of cattle sold individually in a few hours.

Included in that first group of French breeders were: Regis Coudert, Emile Chastanet, Yves Lemaigre-Dubreuil, Pierre Pechdo (general manager of Domaine de Pompadour), Claude Mercy, Paul de Monvallier, Olivier Vandermarcq, Olivier de Rochebouet and Louis de Neuville. Their attendance made our Limousin debut at the National Western that year a big international event, for there was also a large delegation of Canadians and several Mexican cattlemen who came. Even though it was several years before a Limousin class was included in the Stock Show itself, this event attracted a lot of people.

It was Bob Vantrease and his work in promoting show strings of Limousin and setting up scheduled shows around the entire country that brought about the Limousin purebred classes at the National Western several years later.

The year 1971 brought some changes in the cattle industry and affected the Limousin Foundation in several ways. The increasing numbers of Limousin imported into Canada meant that new bulls and new sources of semen were coming into the market. We had planned for semen sales to carry us through the first few years until the income from recordings and transfers would pick up our overhead. Now, several other factors began to influence our financial picture.

The first year of operations, 1968, we had a monopoly on semen from Prince Pompadour through Bov Import. The second year there were six bulls in Canada producing Limousin semen, and we had contracts on three of them, also through Bov Import. They were Dandy, Diplomate and Prince Pompadour.

We were now coming into the critical third year, the year of crisis, according to the business management experts. The need to expand our computer service, the increasing competition in semen sales and a growing problem in collections on semen accounts began to deplete our cash position.

This change took place quite rapidly as the semen situation turned around. That first year we controlled 100 percent of the semen supply, the second year we controlled approximately 50 percent of the available semen in North America, and the third year I estimated we could account from somewhere between 5 percent and 10 percent of the total Limousin semen available. Our semen had all been sold on a direct or mail order basis. Now the commercial breeding centers were coming into the act and they had elaborate sales and distribution organizations. They were cashing in on our early promotion and publicity and doing a good job of out-selling us.

We couldn't complain, of course, because they were building a backlog of future recording applications and memberships. However, it was a difficult interim period, the time between getting out of the semen business and concentrating on membership, recordings, registrations and transfer fees.

To complicate matters, several of our formerly loyal members had ordered large shipments of semen from our three bulls during the past year and suddenly stopped paying their bills. Bred heifers had been selling well and speculation on these cattle had turned a lot of previously conservative cattlemen into varying degrees of "fast buck" traders.

Suddenly we found familiar names of members showing past due accounts of $5,000 to as much as $15,000. There were only a few people involved, but a substantial amount of money was at stake. Most of these had been involved in contract breeding projects and the overhead had gotten beyond them. It meant we had to consider legal action against some members that had been active in building the organization.

In addition, we had contracts on the three bulls with Bov Import that required continuing payments. As the breeding season approached, a further complication arose. Many of our most active members and larger breeders had bought interests in or controlled their own bulls in Canada, thus giving them their own semen supply and putting them in the semen sales business also.

We changed from a position of 50 percent control of semen supply to somewhere around five percent in a very few months. In addition we lost some of our best semen buyers from the previous years. It was a situation I had foreseen, but it happened much sooner than I had expected. Those Limousin bulls were producing semen by the thousands of ampules.

Oxandaburu Ranch, Huron, South Dakota, has a special class at the South Dakota State Fair in 1971 for juniors showing Limousin steers.

For example, Prince Pompadour produced about 10,000 ampules his first year, about 12,000 his second year and the next year his production jumped to 25,000. Two other bulls, Dandy (Bov Import) and Decor (American Breeders Service) each produced an estimated 50,000 ampules that same year, according to our best reports. If other bulls were producing on a comparable basis, the supply could easily approximate 500,000 to 750,000 this year, I told the Board in a June report.

Another PR problem arose out of our success in building a market for Limousin bred heifers. Traders were buying up numbers of open heifers, breeding them to various Limousin bulls, then putting them in feedlots to get fast growth and bloom on these animals.

When they went through the various sale rings around the country they looked good and brought top prices, usually in small lots to smaller breeders who lacked expertise in A.I. breeding. Then, like hot-house flowers that had been "forced," the handling they received produced larger than normal first calves with consequent calving problems on those over-fed heifers.

The same thing had been true, I'm sure, with Simmental and Charolais calves, but we had promoted the easy calving qualities of the Limousin and it blew back in our face. Time eventually showed that with good management and good cows the Limousin were easy calvers, but that summer of '71 did give us some problems.

Our annual meeting that year was moved to October 15 and 16. We had put together a group of true experts on some phases of the cattle industry I felt were especially important to our members at this time.

Our keynote speaker that opening day was Mr. Tom Lasater, the dean of Colorado cattle producers and the originator of the Beefmaster breed of cattle. He was an exponent of using "Nature's way" to manage cattle and made a point of debunking many of the cattle industry's oldest traditions. His key word in cattle management was PRACTICAL!

A panel on the Limousin carcass potential included Cecil Hellbusch, Safeway Stores livestock consultant; Dr. Don Naumann, Food Science and Nutrition professor, University of Missouri; and Bud Prosser, Mead, Colorado.

Selecting and managing a high performance beef herd was the topic for discussion by Martin Jorgenson, famed South Dakota Angus breeder and performance expert. Also on the panel was Dr. Gunther Rahnefeld, director of the Canadian Research Center at Brandon, Manitoba, and the man in charge of the 10 Limousin used for testing by the Canadian government. Louis de Neuville discussed the techniques of sire selection for top performance.

The final panel was on the much discussed topic of bovine blood typing as used in breed association work for sire or dam identification. Dr. Clyde J. Stormont, University of California at Davis, considered the world's foremost authority on the subject, outlined the technique of using blood typing as a bulwark for breeding association integrity. Richard Nelson, of the Holstein-Fresian Association, described the use of blood typing in the Holstein A.I. and registry program. Dr. Dave Bartlett, Vice-president of American Breeders Service, explained the importance of blood typing in any extensive A.I.

Grand champions of the 1971 Great Western Beef Expo was this group of Limousin X Hereford steers. Pictured holding his trophy is J. J. "Bud" Prosser, Bov Import technical director. To his right Kirby Briggs, Kirley, South Dakota breeder. The Dandy-sired group had an average daily gain on the 180-day test of 3.16 pounds per day.

breeding program.

With the advent of widespread use of artificial insemination, the problem of verifying sire data had become more and more important. This became especially true in "contract breeding," where sizable numbers of heifers or cows were bred A.I. over one heat period. This was a fairly common practice in those early years after semen became easily available, often at cut rate prices. It was one way to produce large numbers of halfblood heifers for quick resale. The heifers were generally sold as calves when weaned at a previously arranged price, plus a bonus for the A.I. service. The buyer-contractor usually supplied the semen.

Among established reputable cattlemen this system worked very well. It saved a beginning Limousin breeder from buying a herd of foundation animals, and he could choose a top breeder of good commercial cattle with a performance background. The end result was a hundred or more very uniform calves from a selected bull. In most cases the commercial man liked the contract calves so well he bred back to Limousin the next year for his own use.

Some of these deals, however, did not work out so well and the possibility that cheaper semen had been substituted in the A.I. breeding program involved the Foundation when the Limousin heifers were recorded. Blood typing helped to solve the

problem since all semen producing bulls were required to have their blood typed and on file with both the A.I. center and the Foundation.

As Dr. Stormont explained, blood typing could not always show what bull was the sire, but it could very well prove that a certain bull was NOT the sire claimed. The Holstein-Fresian association had pioneered the use of this technique in A.I. breeding of dairy cattle and we adopted it along with other beef breed organizations that were now into A.I. service. The Charolais association had pioneered the registration of A.I. calves, but the Hereford, Angus and Shorthorn associations had held out against the practice. With the imports of other new breeds where A.I. was the only method of breeding possible because of the Canadian restrictions, the entire A.I. industry boomed.

For this reason our panel on blood typing attracted widespread interest. Few people at that time knew much about the technique, or what its capabilities were. It was a matter of concern to every purebred bull owner who was selling semen and to the breeders who were buying and using semen from different bulls. As higher percentage bulls began to bring higher prices and a few purebred bulls, the progeny of some of the original imported animals in Canada, began to sell in the U.S., the service was a protection for both buyer and seller.

This is characteristic of the French Press coverage we got in those first trips. This was in September 1969 on our second trip to the Limousin. It came about the year the U.S. made one of the first "moon" shots (I think when our rocket first circled the moon). One reporter met us at the train and said, "Why are you here?" I said, "Some Americans had gone to the moon, but we had come to search for the world's best beef cattle!"
—Dick Goff

LA VIE RURALE (Supplement to CENTRE-PRESSE), Limoges, 28 October 1969
He has come from across the Atlantic to import Limousin cattle
M. GOFF: "SOME AMERICANS WANT TO VISIT THE MOON, OTHERS ARE SEARCHING FOR THE BEST BEEF CATTLE"
But the Russians also are coming to buy Limousin animals

"With his wide-brimmed felt hat and his cowboy boots, and in his face something of the Gary Cooper look, he has been seen visiting some of the good breeding herds of our region and reviewing the large gathering of our reddish coloured cattle at Saint Leonard during the cattle show there.

"But he is not here for movies or parades, coming for the second time in a five month interval to the Limousine, M. Goff has explained his mission with a bit of humor.

"Although some Americans are trying to get a better look at the moon, others are seeking a better type of beef animal." M. Goff is one of the latter and he was quick to say that he had already found what he was looking for. His title of Vice President of the North American Limousin Foundation is proof of that. This strong organization already has 115 members with headquarters in Denver, Colorado. It was formed because the American and Canadian breeders wanted to modify the production of beef to give more appeal to the meat consumer, and they believe the genetics of the Limousin breed will help them achieve that goal.

Traditionally, the English breeds, Herefords, Shorthorn and Aberdeen Angus, have been the producers of beef in America and Canada, but the feeling there now is that these breeds tend to produce beef with too much fat for the modern diet.

The North American breeders hope the Limousin crossbreds will produce a carcass animal of slightly larger size, with a faster growth weight and a more flavorful type of meat. In addition, they hope for another advantage from the Limousin and that is the greater ease of calving. This is an important factor in western range country where the birth and survival of the calf is entirely the responsibility of the mother cow, according to M. Goff.

The increase in Limousin exports to Canada and North America is encouraging. From one animal in 1967 it had grown to 16 animals (11 males and five females in 1968) and 79 animals have been purchased by Canadians and Americans this year.

(Translation from the French by Jane M. Goff)

Kansas Colonel, first full French Limousin bull to sell in the United States is pictured in the lobby of the Denver Cosmopolitan Hotel. Shown with Colonel is Bob Purdy, president and Dick Goff, executive vice-president, North American Limousin Foundation; Joseph Hochhausen, president Canadian Limousin Association and breeder of the calf, and Bob Haag, Topeka, Kansas, representing the buyers.

In addition, the first experiments in embryo transplants were beginning with Limousin and the board was asked to approve the registration of purebred calves from this method of propagation. Here again, the decision to approve the registry of "ET" calves was based on the reputation and responsibility of the member, no matter who the technicians were who handled the implant service. But it was a decade or more before this technique was perfected enough to become commercially feasible, and then only on the best purebred animals.

At this meeting, Ron Brown, our systems director, outlined the new refinements in our data handling program and announced the addition of a larger IBM/System 3 computer for in-house handling of our ever growing volume of recording applications. We seemed to run a continuing race between handling capacity and the volume of applications. This too required more money and more staff, but would improve our service to members.

At this meeting new officers and some new directors were elected. They were: president, John D. Moore, Newville, Pennsylvania; vice-president, Burwell M. Bates, Konawa, Oklahoma; secretary,

Fred DeMier, Miami, Oklahoma; treasurer, Bryant Harris, Marfa, Texas; Executive Vice President Richard Goff, Denver; director-executive committee, J. C. Easland, DeSmet, South Dakota. Robert H. Purdy, who had been president since the organization was founded in 1968, became an ex-officio member of the Board.

Shortly after this meeting we enrolled our 1,000th member in the North American Limousin Foundation. We sent him a letter of congratulation and invited him to attend the 1972 National Western Stock Show and opening day Limousin Sale as a guest of the Foundation. He was Darrel Menning of Corsica, South Dakota.

Although current membership numbers now being issued are over 7,500 (1986), that first 1,000 seemed like a major milestone. From that first informal meeting of 10 people in May, 1968, to 1,000 members in 1971, was a sign of some major kind of progress in slightly over three years.

By this time I felt my services to the Foundation were reaching a point where I could go back to building a business of my own. I promised Bob Purdy at the beginning of our organization I would put three years into getting the operation under way. If that wasn't enough, I told him, then it was time for somebody else to take over the job.

Early in December I wrote to the members of the Board and announced I was resigning, effective at

Darrel Menning, Corsica, South Dakota becomes the 1,000th member of the NALF on December 31, 1971. Darrel is shown receiving a new western hat from NALF membership chairman, Cadet Oxandaburu.

1971's TOP SELLING BRED 50% LIMOUSIN HEIFERS

Name of Sale	Calving Date	Seller Address	Buyer Address	Sire	Price
NALF Sale	9-22-69	Dudley-Thompson Hampton, IA	L.F.B. Corp. Grand Junction, CO	Pompadour	$10,100
NALF Sale	9-16-69	Dudley-Thompson Hampton, IA	Gleannloch Farms Spring, TX	Pompadour	7,600
The Garst's	4-30-69	Garst Farm Coon Rapids, IA	H. E. "Buster" Jones Jacksonville, IL	Pompadour	5,500
Oklahoma 71 Sale	4-17-70	Mel Hatley Oklahoma City, OK	John Siebett Hyannis, NE	Pompadour	5,000
Oklahoma 71 Sale	4-16-70	Burwell Bates Konawa, OK	H. E. "Buster" Jones Jacksonville, IL	Pompadour	4,300
Mid-Continent	April 70	Bell-Wood Ltd. Osceola, IA	H. E. "Buster" Jones Jacksonville, IL	Diplomate	3,500
The Garst's	5-8-69	Garst Farms Coon Rapids, IA	McBride Limousin Benalto, Alta., CA	Pompadour	3,500
The Garst's	5-13-69	Garst Farms Coon Rapids, IA	Bell-Wood Osceola, IA	Pompadour	3,300
The Garst's	5-12-69	Garst Farms Coon Rapids, IA	McBride Limousin Winnipeg, Man., CA	Pompadour	3,100
The Garst's	5-10-69	Garst Farms Coon Rapids, IA	Oxandaburu Ranch Huron, SD	Pompadour	3,075
The Garst's	5-19-69	Garst Farms Coon Rapids, IA	McBride Limousin Winnipeg, Man., CA	Pompadour	3,000
The Garst's	5-11-69	Garst Farms Coon Rapids, IA	Bell-Wood Osceola, IA	Pompadour	3,000
Dudley & Thompson	Sept. 69	Dudley & Thompson Hampton, IA	McBride Limousin Winnipeg, Man., CA	Pompadour	3,000
Oklahoma 71 Sale	4-19-70	Burwell Bates Konawa, OK	Limousin Cattle Co. Pauls Valley, OK	Pompadour	3,000
The Garst's	5-13-69	Garst Farms Coon Rapids, IA	Casey Anderson Calgary, Alta., CA	Pompadour	2,900
The Garst's	5-27-69	Garst Farms Coon Rapids, IA	Baldo, Inc. North Platte, NE	Pompadour	2,800
The Garst's	10-1-69	Garst Farms Coon Rapids, IA	Oxandaburu Ranch Huron, SD	Pompadour	2,750
NALF Sale	5-13-69	Geesen Enterprises Agate, CO	McBride Limousin Winnipeg, Man., CA	Pompadour	2,750
NALF Sale	6-21-69	Bob Purdy Buffalo, WY	Wm. Schermer & Sons Latimer, IA	Pompadour	2,750
NALF Sale	5-5-69	Geesen Enterprises Agate, CO	J. V. Elliot Fresno, CA	Pompadour	2,750

The first major steer championship came to the breed when both the grand champion and reserve grand champion steers at the 269-head 4-H show held at the 1971 Calgary Stampede were Limousin crosses. Both steers were bred by LK Ranches, Bassano, Alberta and were by Decor. Kirk Gosling, Dalemead, Alberta, the 14-year-old exhibitor, is pictured at the halter of the grand champion.

the end of January, 1972, after the January National Western Stock Show and our second annual Limousin cattle sale was over. We put a great deal of effort into planning and promoting this sale, under Jim Baldridge's direction, and this event would see the first full French Limousin bulls offered for sale in the United States.

The committee consisting of Jim, Charles L. Moore, Bruce Waddle, Louis Swift and myself, put a lot of time, work and planning into the project. Jim and Charlie had personally traveled over much of the country to see and approve most of the consignments.

At high noon on opening day of the 1972 National Western Stock Show, Friday, January 14, 1972, Auctioneer Jim Baldridge began taking bids on the first full-French purebred Limousin bull to sell at auction in the United States.

He brought a top price of $56,000 and the buyer was Robert Haag of Topeka, Kansas. The eight-month-old calf, named "Computer," was a son of Eclaireur, out of a purebred Beneyton female,

Eponge. He was bred and consigned by Wilfred Hochhausen, Edmonton, Alberta, Canada.

A second purebred bull calf named Challenger, only three months old, brought $40,000 from Curtiss Breeding Service, Cary, Illinois. He was sired by Dard, a half-brother to Prince Pompadour, and out of a Chastanet-bred female, Elfe. The calf was consigned by XIM Genetics, Guelph, Ontario, Canada. These two calves meant the advent of the first available sale progeny from the French imports into Canada a few years before.

They were the vanguard of a continuing line of purebred Limousin cattle that sold at the National Western over the coming years for ever-increasing prices.

The crossbred cattle in the sale were a select group of carefully sifted animals including a three-way pair of a half-Limousin cow by Prince Pompadour bred back to Dandy, and with a Decor three-quarter calf at side. The consignment was from Oxandaburu Ranch in South Dakota, and was bought for $15,000 by Mr. and Mrs. Rick Dobson and Dr. and Mrs.

President Bob Purdy is pictured with Gloria Jennings, Highmore, South Dakota, the first Limousin queen in the United States.

Glenn Willis, both of Oklahoma.

Another similar package of a halfblood Pompadour cow, bred to Dandy and with a three-quarter heifer calf by Echo, sold to Lundvall Bros. of Greeley, Colorado. The pair was consigned by Woodbridge and Son of Iowa.

The three-quarter Limousin open heifers averaged $4,800 each with the top animal going to J. V. Elliott of Rancho McLin, California, for $5,700. Next highest three-quarter heifer sold to H. M. Jordan, Mississippi, for $5,400.

The 18 bred heifers averaged $2,250, and 52 open heifers averaged $795. J. V. Elliott was the top buyer in the sale, paying a total of $47,300 for 23 open and bred heifers. Second highest buyer was a Budrick Farms-Bon Vita Land and Cattle combination of Oklahoma whose bids totalled $20,225. Third ranking buyer was Floyd McGown of Texas who bought a select group of three-quarter bred and open heifers for $11,125.

As it had been the year before, the sale barn at the yards was completely packed. It was standing room only at every square foot of open space in the seating section. The Limousin sales had already become a tradition that attracted tremendous crowds. Although a lot of old-timers in the business were skeptical of the breed, they came to the sales and they were impressed, both with the cattle and the prices.

My resignation in December was not completely unexpected because I told the group at the last Board meeting that I would like to get back into freelance promotion work on my own. The Foundation was entering a new phase of operations and I felt the initial work was done.

My old friend Farrington Carpenter summed it up one time in a discussion on this subject. "Sometimes, Dick," he remarked, "the greatest contribution anyone can ever make to a major project of this kind is to get it started."

As far as the North American Limousin Foundation was concerned I felt I had gotten it started, I had conceived the idea many years before, organized it and put a lot of my own time and money into the effort long before the first organization meeting was held.

Thanks to Adrien de Moustier, Bob Purdy, Charlie Moore, Bruce Waddle, Sherman Ewing, Jim Scott and Bill Smutz, the ground work was laid and

Retired Navy Commander, John Moore, Newville, PA was the second NALF president. A former Charolais breeder, John was active in promoting the breed in the northeast through his Keystone herd, the Pennsylvania Keystone Livestock Show and the Eastern National.

the initial concept developed. Working behind the scenes, and unknown to most of us at the time, was Louis de Neuville. His support and guidance over the forthcoming years were a major contribution to the over-all growth and success of the Limousin breed in both Canada and the United States.

His appearance at one of our most critical meetings during that first year gave our movement an initial impetus that has never diminished. Like the ancient Greek's idea of the "deus ex machina" influence upon the play of life, the centuries-old concept of the French Limousin cattle and their pragmatic breeders were shaping the economics and the destiny of numerous individuals in the livestock industry over many parts of the world.

We were fortunate in those early days to have the unstinting help and support of some exceptional Founder members. Bob Purdy was a major influence at the beginning as our first president. He had been president of the American International Charolais Association when it first adopted the practice of registering A.I. calves. He had done an outstanding job of promoting his Charolais herd and when he announced the comparative weights of his first Limousin-cross calves from his Charolais cows a lot of people were impressed.

The cutability report on his "909" bull aroused a great deal of interest among many large cattle operations that he knew from his contacts in the American National Cattlemens Association. He was an outstanding businessman and had a world of corporate management experience behind him.

In addition there were a number of scientists and technical people who were helpful with their time and advice when it came to questions of genetics and factors of performance in planning our program. One of those who gave us a great deal of time and experience was Dr. Robert de Baca, then of Iowa State University.

Dr. Carroll Schoonover, University of Wyoming; Dr. Gunther Rahnefeld, Brandon Research Center, Canada; Dr. Tom Sutherland and Dr. Jim Brinks of Colorado State University, all attended our early meetings when asked and offered suggestions and advice, often spending long hours in committee meetings to help shape the basic ideas of a performance oriented breed organization.

Then too, there were all those individual members who worked long hours on committees and sat through endless discussions to shape our first set of rules and regulations and then worked out numerous revisions as new suggestions were made and adopted. Not half of them are listed among the early officers and directors of those first three years, but time and space limit the listing of their names. We can be grateful for all their unselfish efforts and the benefit of their experience.

After the news release on my leaving the Limousin Foundation I had a number of phone calls and letters from friends and well-wishers. One of these in particular summed up the situation among many members of the livestock press. I considered it more of an accolade for the Limousin Foundation than it was for me.

It was from Frank Lessiter, editor of the *National Live Stock Producer* magazine in Chicago. He said:

"Dear Dick: I was quite surprised this morning to read the news release that you are leaving the Limousin Foundation. I wish you luck in your new projects.

"You have done a great deal for both the Limousin Foundation and the entire U.S. beef industry in the last four years. From where I sit, it looks like the Limousin Foundation is being more progressive and doing more things right than practically any other breed organization in the country today.

"Hope to see you during the American National Cattlemens meeting in Denver next month. Sincerely yours, Frank."

Perhaps all of us working together made some small contribution to the beef industry. Certainly those who followed and developed the present sire summary and the data bank of herd performance data, have achieved the goals we only hoped for in the beginning.

In my last newsletter I told the Board of Directors I felt the breed was destined to play a major part in the future development of the American beef industry. It was called at one time by a Canadian animal scientist the most efficient beef producer in the world. The proof, I said, may be slow to come, but I sincerely believe it will eventually become known as that type of animal.

Dick Goff shot this picture of a cow and her big bull calf at Louis de Neuville's farm on his visit to France in 1969. An accomplished photographer, Dick's file of slides and prints made on his early trips to France supplied the graphics for all of the early breed promotion in the United States and Canada.

Part Two 1972-1978
The Development of the Breed

by Dale F. Runnion

Chapter 1
1972—A Year of Change

It was time for the Limousin breed of cattle to take its place among the other breeds in the United States. The cattle were performing satisfactorily for everyone who had them. The genetic base for Limousin semen had increased from one Limousin bull to 70. The fullblood females of Canada were producing calves and in January 1972 the first two fullblood calves were sold in the National Sale in Denver to U.S. buyers. The gates were opening for pure French cattle ownership by U.S. breeders.

The resignation of Dick Goff as executive vice president of the foundation in January 1971 created a big hole in the breed's leadership chain. The fact that we were going into what appeared the most exciting year for major breed expansion with less than full power in our leadership spot concerned the board and our membership.

The demand for the new European breeds was here and it was real.

The Simmental breed had come into Canada just one year ahead of our Prince Pompadour. The American cattleman was chomping-at-the-bit for something different; Simmental seemed to fit the niche wanted by our cattlemen and advocated by many of the academia. The breed had more growth and more milk than their English contemporaries. People liked Simmental because they were structurally sound.

It seemed that for every heifer we sold for $10,000, the Simmental breed would sell 30 for $30,000. A

Burwell Bates, Konawa, Oklahoma, 1972-73 NALF president, pictured with William Schermer, Latimer, Iowa. Both served on the NALF board and as presidents of their respective state associations. Both actively promoted the breed at home and with a show string on the road.

Limousin Journal *publisher Dale F. Runnion was the second Executive Vice President of the North American Limousin Foundation. A graduate of Ohio State University, Runnion was general manager of the* Angus Journal *and later a partner in the sale management firm National Livestock Brokers. He took leave from the* Limousin Journal *to serve on an interim basis for the NALF.*

lot of people were making money and a lot of people were on the road saying "buy Simmental."

At the January 1972 board meeting I was hired as executive vice president and given the added duties of chairing a committee to conduct a search for a permanent executive vice president. I asked the board to appoint John Moore and Burwell Bates, current president and vice-president of the NALF, to the committee. They honored my request and gave me a year to complete the assignment or until I found the person I thought would be a suitable replacement.

At the time I was editor and publisher of the *International Limousin Journal* in Fort Collins, Colorado. In February I started my daily trips to the Denver NALF office or to some other center of Limousin action in the United States or Canada. Dean Jacobs and my wife June took charge of the breed magazine in my absence.

The backlog of paperwork in the NALF office topped my list of priorities. An influx of registrations came in following the '71 fall weaning, producing huge files of applications for processing. Requirements for registration eligibility were changed and former electronic systems were unable to handle the necessary storage load. A new system was to be delivered in January. It had to be programmed before production could start. Meanwhile applications continued to pour in to the NALF office and the files already bulged with unfinished work.

The office staff was working fulltime and over-time besides, logging in the new applications, digging out and hand producing certificates needed for sales or transfers. I organized a special crew and started them hand producing certificates and had mailed out 3000 papers before the new IBM system came on-line.

In March an ad was prepared for the seven major multi-breed magazines and weeklies. "Limousin Crosses Win at the Bank" featured Jerry Adamson and his grand champion carload of feeder calves at the 1972 National Western Stock Show.

On March 22 the selection committee unanimously recommended Robert H. Vantrease to the board for the position of executive vice president from an impressive list of applicants. The board concurred and Bob was hired to assume his duties April 1, 1972.

Vantrease was raised in Lebanon, Tennessee. He is a graduate of the University of Tennessee where he was active in campus activities. He was a member of that institution's Livestock Judging team that won the International Contest in 1962. The experience gained as vice president of the Tennessee State 4-H club and many other college organizations would serve him well as he was called upon to deliver the Limousin word in the coming years.

While in college and after graduation, Vantrease worked for three purebred Angus operations before he was appointed the *Angus Journal* fieldman for eight southeastern states. From this position he moved on to become director of purebred sales for Premier Corporation in Michigan. Because most of his sales contacts were with western ranchers, he traveled out of Denver, Colorado.

"My first sincere interest in Limousin came on one of Premier's ranches near Roggen, Colorado when I saw the calves on that ranch by both Limousin and Simmental bulls out of Angus cows. Some 400 head of these Limousin steers were put on feed. I took pictures of them; I was impressed with the uniformity and their muscularity. As the feeding period went on they didn't slack off on feed or gain. When they were marketed just short of 1200 pounds they were still going on.

"They hit a home run in the packing house scoring high in yield and quality grades and rated accolades from the packer with their cutability and noticeable lack of excess fat," Vantrease said.

The move to new Continental breeds (most called them exotics) in the United States was going on at a fast pace when Bob Vantrease took over as executive vice president. The NALF had recorded 30,000 Limousin percentage cattle by that time as compared to more than twice that figure of registered Simmental in the U.S. The first Chianina semen was coming into the U.S. from both Canada and Italy. There were a number of other new breeds on the way to Canada. The mood of the cattleman in the U.S. was "more milk and bigger" and to get there as fast as he could.

The story that circulated around the country at that time relates a visit behind the barn with a northwestern U.S. farmer who proudly pointed to a pile of frozen calves behind the barn saying, "You know, I'll bet half of those calves weighed over 130 pounds." Thankfully, they were not Limousin crossbreds.

At this time it appeared the well conceived plan in founding the NALF had two factors that now were proving to be detriments to the expansion of the organization: performance requirements for registration and in-house semen sales.

The semen business had created huge debts for the Foundation by a number of breeders and operators who let their enthusiasm for the breed and the desire for the fast buck extend their expectations beyond the realm of good judgement and management. A big share of the time at early board meetings

was spent with this problem. The advent of new Limousin sires left the Foundation with a semen inventory of over 8,900 vials. The accounts receivable, computer and organizational costs had drained the operating capital to a very low figure.

Limousin semen sales were going well all over the country, but at this time there was not only a backlog in the office of unprocessed registration applications and transfers, but an even bigger number of half-blood Limousin cattle in the country not yet recorded that were over 15 months of age. The reason for the breeders' reluctance to register these cattle seemed to be their inexperience in the registered cattle business and the complicated forms requiring performance data for registration. The forms called for data most ranchers were not accustomed to keeping. Vantrease and the board moved to correct the situation. Vantrease says, "The recording procedure was complicated. It called for first applying for registration, receiving a temporary registration paper

Don O'Brien, Pineville, Missouri displays a premier breeder trophy at the Kansas City Royal. Don was a Founder member, served on the national board, and was active in three state associations—Oklahoma, Arkansas, and was president of Missouri.

That Man Vantrease!

By Harald Gunderson

Robert H. Vantrease

Some guy that Bob Vantrease!
Part executive. Part cowman. Partly showman and partly promoter. And all Limousin. Through and through.
It's a real education to get to know this man who serves as chief administrator for the North American Limousin Foundation out of Denver, Colorado. Know him better and you'll like him more.

Harald Gunderson, founder, publisher and editor of the *Limousin Leader,* Calgary, Alberta, writes about the NALF executive vice president Robert H. Vantrease. Vantrease was in Canada to speak at an agricultural seminar of the annual meeting of the Institute of Chartered Accountants of Alberta. After this speech, he addressed the board of directors of the Canadian Limousin Association and visited several Canadian herds and bull studs. Gunderson wrote of Vantrease after his trip to Canada:

...Wherever he went, whatever he did, the message was the same. Limousin is coming on strong. Get out of the way, people, and let those cattle go! Got a problem? Heck, man, right's right, and wrong's wrong...

Square shoulders rest on a big frame, a Stetson or a summer straw identifies this man in motion, and a boyish smile breaks over a face that is often sombre or set deep in thought.

A super Limousin salesman. A man who got his education in the halls of learning and the conflicts of life. One-time ringman at sales and field representative for the *American Aberdeen Angus Journal.* He's nobody's fool and nobody's doormat. Just mark him as Mr. Believer in Limousin...

That's the sound of the Vantrease enthusiasm and the CLA board of directors got a good bellyful of it as Big Bob let go. It was good medicine for the Canadians and served to lift their spirits at a time when association problems, a closed border and sagging markets were causing even the best to lose heart.

Down in the dumps with exotics? Taking the temperature of Limousin? Got the banker blues? Just call in Dr. Bob. You'll be glad you did.

"We talk about Limousin that go back 7,000 years in French history. You want to know something? We've found out more about these cattle in seven years then the French discovered in 7,000 years. Now the French are performance testing and that's got to be a great sign."

A man of deep loyalties and sense of obligation to his employers, Vantrease cautions that performance testing and research isn't the total answer. "If it was, blind people could select cattle." But, he quickly adds, "we've spent a lot of money on research to find how little we really know."...

"If you select a poor bull you deserve what you get."

Friendly and frank, open and approachable, Robert Vantrease is a great ambassador for his country and a damned good man to have around.—*Limousin Leader.*

---

back from NALF, then re-submitting the paper to the Foundation after yearling weights were taken and a permanent certificate issued to the breeder. The program was too idealistic for practical breed growth.

"The board reasoned that only a change in the requirements for registration would give us a base that would put our breed in a position to command a larger market share so we could continue to expand our breed promotion efforts. We knew the interest in Limousin was for real out there in the country because semen sales were good. These over-age halfblood females needed to be registered to increase our genetic base for 75 percent offspring.

"The board decided to accomplish this with two moves. First, the breeder would have the option of whether he wanted to register his calves with or without the performance data. Secondly, breeders would be given the opportunity to register all halfbloods that were over-age at no penalty. The registration of these over-age cattle produced some $200,000 in revenue and brought 26,000 registrations into the herd book."

Limousin cattle started 1972 with a most successful sale held in conjunction with the National Western Stock Show. The "standing room only" crowd in the huge sale pavilion was indicative of the interest in our breed.

Another barometer that the cattle were doing their part in attracting the cattleman's eye was the feeder cattle show in the yards at that 1972 show. There a load (15 head) of Limousin-Angus cross steer feeder calves were judged reserve champion crossbreds of the show. The calves were produced by veteran cattleman Jerry Adamson, Cody, Nebraska. The 560 lb. calves were out of first calf commercial Angus heifers calving as two's to the service of Diplomate and Dandy.

They were a popular load in the yards and when they came into the ring the next day they set a National Western Feeder Show auction record when they averaged $1.35 a pound ($769.00 a head). The top calf brought $3.30 a pound and totaled $1,782. They outsold the grand champion load of the show (straight bred Ankony Angus calves) by 63 cents a pound.

Close on the heels of the Limousin successes of the National Western came the Great Western Beef Expo. It was the first national event attended by Vantrease as the breed's executive vice president. Sire groups of five steers each were entered into a contest measuring weight per day of age, carcass weight per day of age and pounds of quality red meat per day of age.

There were 49 sire groups (145 head) in the contest. Ten of the sire groups were Limousin.

The Bov import sire group by Dandy (X Noel, bred by Chastanet) was declared the supreme champion at the show. They won the supreme award after topping the pounds of live steer per day of age, carcass weight per day of age and pounds of quality red meat per day of age divisions of the show. Weights in all three categories were new Expo records. The 433 day old Dandy steers produced carcasses averaging 787 pounds with a yield grade of 2.4 and they sported 15 inch ribeyes.

In this contest another Bov Import sire, Diplomate (X Alaska, bred by Bourbon) sired a steer that produced the highest lean yield per day of age (closely trimmed, boneless lean from primal cuts) of .944 pounds per day of age over the 245 steers in the show. All the Bov entered steers in the contest were bred by Hoyt Nicholas, Ree Heights, South Dakota.

Bob Vantrease has always felt that winning this contest before the eyes of so many commercial cattlemen was one of the breed's greatest triumphs toward breed recognition.

Robert J. Steward, Keating, Oregon was Founder member #71. Bob was an active member of the Oregon and other west coast cattlemens associations as well as the National Cattlemans Association. He was a NALF board member and was instrumental in organizing the Western States Limousin Association.

I have always maintained the purebred livestock business was a "people business." In this era of our breed expansion it was "the people" that began to show their influence.

The Limousin breed was attracting the right kind of cattle people. There was no question about it. Friends I had made during my association with other cattle breeds commented on this to me many times. There was an "air" of family whenever Limousin events were held. Those of us working in the breed all the time were certainly aware of this atmosphere and it was refreshing to know that it was noticeable as well to the folks involved in other breeds.

The quality kind of people and their dedication to the breed was just the edge we needed. We were attracting substantial, influential farmers and ranchers and businessmen. They were the kind of people who were successful in their own businesses but they were also leaders in their respective communities and states. This was a big factor in influencing their peers and friends to get on the Limousin bandwagon and become breeders of Limousin themselves, thus bringing more new members to the NALF.

The "good people" influence started with the board of directors who were responsible for establishing the breed direction and policies. They served and acted unselfishly and wisely.

Merchandising a breeder's product to the best advantage should always be the top priority for a livestock breed. The Limousin breed was fortunate that it had a strong breed magazine *(Limousin Journal)* with a knowledgeable experienced staff. They worked closely with the NALF board in mat-

C.K. "Sonny" Booth has auctioned more dollars worth of Limousin cattle than anyone in the world. He is at home in Miami, Oklahoma and had much to do in making it the most concentrated area of Limousin cattle and number of breeders in North America.

Ken Holloway, Chattanooga, Oklahoma is the senior partner of the sale management firm of American Cattle Services. His total involvement with Limousin through his own successful breeding herd and his management firm has contributed greatly to the breed's advancement. Fred DeMier, Miami, Oklahoma (pictured left) gave Ken his first Limousin sale assignment.

ters of policy. They served a very worthwhile purpose in communicating and endorsing those policies to the membership. The magazine received the major portion of its revenue from advertisements promoting sales. Having a good sale was therefore important to the *Limousin Journal* as well as to the breeders who were having the sales.

Sonny Booth was a young auctioneer born and raised near Miami, Oklahoma who got the nod to auction some of our breed's early sales. To this very day he handles the duties of auctioneer at most of the Limousin auction sales in the United States and Canada. His skill in the auction block, his dedication to the breed and the personable way he conducts his business is appreciated by the consignors and buyers alike. His full sale schedule is a true indication of the confidence breeders place in his ability to do the job right for them.

Sonny recalls some of his early experiences with Limousin and why people switched to this particular breed. "The Oklahoma area was fascinated by the Limousin when they bred their cattle to them and got a calf weaning at six hundred and fifty pounds. This was a miracle, because anyone could have a five hundred pound calf. It was a two-time winner with an easy delivery and a big calf at weaning. Breeders from the sandhills of Nebraska thought it was a great deal to have an eight hundred and fifty to nine hundred pound yearling, as they were used to seven hundred pound yearlings."

It was this kind of "Booth thinking" that helped make northeastern Oklahoma have the largest concentration of Limousin in the United States.

Ken Holloway, Chattanooga, Oklahoma, like Sonny, an Oklahoma State University graduate, hung out his shingle in the early beginnings as a sale manager. An astute young man, experienced beyond his years, Ken became not only an able administrator in the conduct of a successful auction sale but a confidant in the herd management and breeding programs for breeders and investors all across the country. Like Sonny, he is a keen judge of livestock. The opinion of both these men is respected and acknowledged by folks involved in the breed.

In 1971 there were scores of halfblood heifers running with their dams in contract herds in South Dakota. It was a time when private treaty sales action was in limbo. "What are they worth" was the question behind the impasse that existed between buyer and seller.

In late summer of that year Ken Holloway and Sonny Booth made a trip to South Dakota and purchased over 400 head of these 50 percent Limousin heifer calves. Some they had orders for, others they purchased for their own speculation and herds. It started the trade moving.

In the early days of our breed many breeders were wanting to have auction sales but they had no ex-

Bates Limousin Ranch, Konawa, Oklahoma has the distinction of showing the first Limousin champion in the United States. Popo Dora (x Prince Pompadour) was a halfblood bred to Diplomate and sold for $3,000 in the 1972 Oklahoma Limousin Association show and sale.

perience whatsoever in staging such an event. Ken would make a visit to the farm, select the saleable cattle, advise and set up a feeding and management program and pick up the pedigrees, catalog and advertising material. This was standard practice for sale managers of that period.

Before sale day arrived Ken would appear at the farm or sale site with a crew of men. They arrived with squeeze and blocking chutes and would clip, wash, tattoo or whatever else needed to be done to put the offering into sale shape. Many young men who have since graduated from Oklahoma State University and are now breeders or have assumed responsible positions in agri-business, got their first taste of hard work in the chutes, sorting cattle or digging post holes for display pens for the coming sale as a working member of Ken's crew.

This extra touch soon gave buyers an appreciation of the fact that cattle bought from one of Ken's auction sales would have the paper work in order. The seller could be proud of the cattle because they were presented in a professional manner. The process helped immensely in the creation of a price base for Limousin percentage cattle. It has been said that an animal is only worth what a buyer will pay for it. The Holloway plan and crew made the buyer more aware of the quality in the sale ring.

In 1971 Ken and his American Cattle Services organized the first show and sale for the breed. The Sale of 1972 was planned and sponsored by ACS in cooperation with the Oklahoma Association. The distinction of showing the first Limousin champion in North America went to Bates Limousin Ranch, Konawa, Oklahoma.

Ken and his organization created many firsts that combined a show with a national or regional sale. They founded the All American Futurity show and sale, the Louisville National, the Fort Worth sale and many other events that have been repeated year after year. Ken was, and still is, a strong believer that the march for the purple creates goals and thrills not attainable in the livestock business anywhere else except in the show ring.

1972 had indeed been a year of change. A new executive vice president was appointed. Membership doubled during the year and our base of available pure French Limousin bulls was increased to 27 head. Registrations of percentage cattle reached 39,000. The Foundation was in a strong financial position as working capital had doubled within the year.

The market for Limousin and percentage cattle was strong. During the year 18 full French bulls sold at auction in the U.S. and Canada from $13,500 to $176,000 to average $38,473. The nine top selling purebred females averaged $21,858; 26 top selling 75 percent females averaged $6,088 and the 26 top selling halfblood females went for $4,670.

The breed position was strong, but it needed to be because the U.S. cow herd had been steadily increasing for several years. The next four years we faced a depressed commercial cattle market in the U.S. that only a breed in a strong position would have a chance to survive.

Harbinger was the first fullblood female to sell in the United States. Imported from England by Raymond Hefner, IBSI, Oklahoma City, Oklahoma (pictured left), she sold to Budrick Farms, Mannsville, Oklahoma for $50,000 at the Fort Worth sale.

Chapter 2
The Search For Breed Advantage

In spite of a great many Limousin successes in the show ring and in the feed yards in the early 70's, there was still opposition to our breed on many fronts. "This new breed was different," Bruce Brooks, Springer, Oklahoma auctioneer and Limousin breeder says. "I think the biggest problem early Limousin breeders had in merchandising their product was to make commercial producers accept the Limousin pattern and the amount of muscle they carried. Keep in mind that the early 70's saw the first Limousin in the United States, but it was also a period in which the Charolais cattle were losing so much popularity with the commercial cattlemen because of their extreme muscle pattern and the bone of these imported sires, thus causing a great deal of calving problems. With time and promotion and a certain amount of selection, the Limousin breeders were able to sell the commercial breeder a stout, well-muscled bull that would cover the ground aggressively and sire small calves at birth that hit the ground and weaned heavy.

"Feet and leg problems were also an early problem with the breed, but once again, through selection and planned matings, the Limousin breed did survive and conquer this problem."

At this time in our development, the Foundation and breeders were constantly in a search for facts that would spell out even the smallest amount of breed superiority or advantage. We were looking at test station results, national experiments, the show ring and breeders and feeders testimonials to justify our own breed dedication and produce selling points for our peers and breeders.

The win for Limousin at the Great Western Beef Expo was heavily promoted by both Bov and the Foundation. It was our way of saying, "Limousin might not be first in yearling weight but when you take their hide off you have 16 pounds more quality red meat for sale than the other two big beef breeds."

Much of the early performance testing was being done on feed rations adopted for the English breeds in this country. It was a ration that was programmed for Angus and Hereford cattle to finish at 1,100

The Supreme Champion Market Steer at the 1972 Western Stock Show, Edmonton, Alberta was a Prairie Pride son out of a Shorthorn cow. The 980 pound champion was the only Limousin cross entry in the 239 head show. He was bred by William Hebson, Okotoks, and shown by Lynn Armistead, Onoway, Alberta.

pounds and grade Choice.

Our Limousin crosses would take a hotter feed. Our observations of the Premier cattle on feed in Fort Collins, Colorado told us that you could wean Limousin crossbreds off the cow, work them into a hot corn ration in a short time and they would finish and grade at 1200# and yield grade 90 percent ones, twos and threes. It was an advantage our breed had but if the ration was not there our Limousin did not finish.

At the Brandon Research Station, Manitoba, Canada, Limousin and Simmental steer and bull calves were put on feed test. The Limousin steers and bulls were tops in feed conversion, had one square inch larger rib eyes, had high dressing percentage and calved easier than their Simmental counterparts. Seventy-two percent of those steers killed at 1,000 pound live weight graded U.S. Choice and 100 percent of those killed at 1,200 pounds made the Choice grade. The test director, Dr. Z. R. Rahnefeld commented at the field day, "The Limousin appears to be a much more efficient growing animal than the Simmental."

The results of the test on the 1972 calf crop at Brandon were essentially the same. Limousin were easier calving, more efficient and graded better.

Data collected at the U.S. Animal Research Station at Clay Center, Nebraska on their comprehensive germ plasma evaluation program in the first two

Prof. F.S. "Sloan" Baker, University of Florida, declared the Limousin crosses in his test "the most efficient I've worked with in 20 years." He found the USDA grading charts did not encompass figures to include the meaty Limousin crossbreds with their high cutability and large ribeyes.

Ben and Ruth Price, Reading, Kansas started in the Limousin business with Founder membership number 12. They had one of the first sales in the United States featuring Limousin. Both actively worked with the Kansas Association. Ruth served two terms on the NALF board.

tests gave Limousin cross steers advantages in feed conversion and carcass traits.

Unfortunately, I do not think a big percentage of the commercial cow and calf men were listening when we related the advantages of Limousin. For several years the rancher had been told the importance of weaning weight in calculating his profit. I think they were looking at weaning and yearling weights and missing the Limousin advantages. Many cow and calf operators associated our full French bulls in their minds with problems their neighbor breeders had encountered with imported Charolais. I also think that one of the real problems in corralling the cow and calf men was the fact there was no one willing to pay a premium for the feeder calves they produced even though we were saying they were more efficient in the feedlot.

It would take time until we could get 75 percent Limousin bulls out on the range with the cows and more feeders through the feedlot and tested by the packer before a meaningful expansion into the commercial cow and calf business would be forthcoming.

Jim McBride, High River, Alberta, Canada says, "The Limousin business was being kept alive by the Limousin winnings in the carcass, feeder and market steer shows. Eight or ten years of winning carcass contests made the other breeds sit up and take a note. When the cow and calf man found out Limousin calved easier our breed moved."

In a trial at the Agriculture Research and Education Center (AREC), University of Florida, a group of 75 percent Limousin steers demonstrated their ability to gain rapidly, continuously and efficiently from weaning to slaughter and to produce carcasses with 19 percent more edible beef than expected.

The average gain of the steers was 668 pounds (2.94 pounds daily) and the feed conversion or feed per pound of gain was 5.38 to 1. This attests to the fast, continuous and highly efficient gain made by the calves. Professor F.S. "Sloan" Baker monitored the test.

When cattle carcasses are yield graded, the carcasses are evaluated on cutability or estimated yield of closely-trimmed boneless cuts from the chuck, rib, loin, rump and round. This estimated yield of preferred cuts, which is largely edible meat, can be calculated on a percentage basis from the thickness of fat over the ribeye, the area of ribeye, the percentage of kidney-heart-pelvic fat and the carcass weight by a formula worked out by the USDA grading service.

Using the USDA formula, which was developed from data on British breed carcasses, the yield of preferred cuts for the three-quarter Limousin carcasses was estimated to be 51.8 percent. When the carcasses were actually cut and trimmed to 0.5 inch fat cover (the normal fat cover for finished cattle) at the Florida Animal Science Meat Laboratory, the yield of preferred cuts was found to be 61.7 percent, which was 19 percent or 65 pounds more closely-trimmed cuts than predicted by the USDA formula.

Thus, instead of yield grade 2 carcasses, actual cutout revealed that the yield of the three-quarter Limousin carcasses was too high even for the charts. These steers yielded from 52.4 to 54.6 percent of closely-trimmed preferred cuts.

It was apparent that data was needed to develop a formula to accurately estimate cutout of high yielding cattle such as these three-quarter Limousin crosses. If the extra 65 pounds of closely-trimmed cuts in the three-quarter Limousin carcasses was worth $1.50 per pound, each carcass would have cut out $97.50 more meat than estimated by the USDA formula.

As previously indicated, data was being accumulated at the University of Florida Animal Science Meats Laboratory to assist in developing a formula for high yielding cattle. The meatiness of the 75 percent Limousin carcasses, especially in the tremendous rounds, was not taken into consideration by the regular USDA procedure for estimating cutability, which as previously stated, was worked out with British breed carcasses.

From this small group of three-quarter Limousin cross calves, we obtained data that indicated the possibility of obtaining fast-gaining, highly-efficient feeder cattle that yielded a much greater proportion of preferred cuts than previously thought possible.

On-going research the following year in the AREC program compared two sets of Limousin steer calves on the basis of deferred feeding versus the full feedlot after weaning. The results of the comparison showed a distinct advantage in taking Limousin calves from weaning direct to the feedlot and on through to market weight. Advantages were in terms of less time to reach market weight and in efficiency of feed conversion. Cost of gain was substantially less for the full-feed steers compared to those that ran on pasture before entering the feedlot...a concept some feeders might find hard to accept.

The calves used in the study were cull bulls from the Barry Ottinger herd near Quincy and were certainly not "selected" for the project.

"These Limousin have a great advantage in feed efficiency and carcass cutability," according to Baker, "plus they're a moderate sized animal. People are beginning to learn that the biggest isn't necessarily the most efficient."

Feed cost per 100 pounds of gain was $3.06 in favor of the steers that went on feed directly after weaning. These feedlot calves were fed 183 days to reach the 980 pound kill weight while the deferred feeding steers spent 213 days (112 days pasture and 101 days in the feedlot) to reach 885 pounds (weighed 22 pounds more at the start). Both groups dressed 62 percent.

The Florida test results were widely circulated to the industry by the Foundation.

This search for the advantages our breed had to offer the commercial cattleman led Dean Jacobs and me to organize the Limousin Bonanza. The show would be held in July 1973 but we started on the plans in March 1972. Our concept was to have a steer show that would be judged solely on carcass value per day of age.

Four cattlemen meet at the Noyes Feedlot, Minden, Nebraska to plan the 1973 Bonanza II. Pictured left is Marfa, Texas cattleman Bryant Harris. He was the 1973-74 NALF president and Founder member #93. Bud Prosser, Mead, Colorado had a hand in most every important promoticn event involving live or carcass cattle in the formative years. Carlton Noyes served two eventful years as NALF President. Bob Vantrease is pictured right.

Jack Eberspacher, Beaver Crossing, Nebraska, a 4-H member, took his 1,120 pound Limousin x Angus steer to the grand championship spot in the 864 head Ak-Sar-Ben 1972 show. The steer was by Decor and sold for a show record dollar top of $11,150.

The breed's first steers were hitting the show circuit and we wondered if we would ever have a Limousin grand champion. We knew if we had a show of our own we would be sure of a champion by a Limousin.

Our fears of having Limousin champions at other shows were not warranted. In the summer and fall of 1972 the grand champion of the 864 head Ak Sar Ben show, the grand and reserve champion at Bakersfield, the grand at the Ozark Empire, Pacific National and Calgary were all Limousin. The carcass champion at the American Royal, Sioux Falls and many other shows were wearing Limousin hides that same year.

The Bonanza show had to be different. People had to see what tremendous carcasses this breed could produce. Many of our breeders had never seen a 15 inch rib eye; the breed needed national exposure. A record purse, great participation and a unique judging system was planned to make it so.

I raised a $10,000.00 purse by making 20 calls to breeders for their support. Louis de Neuville and Cofranimex sent elegant gold engraved Limoges porcelain vases for the champions.

The show was designed to combine performance and carcass merit into retail value per day of age as a final judging standard. Steers were nominated in the futurity-type show in November, 1972. Entry forms recorded sire, tattoo, herd prefix, birth date, breed of dam, weaning weight, weaning date and color markings by the breeder and owner at the time of nomination. After nomination, each steer was tagged with a special "Bonanza" ear tag to wear until slaughtered. Steers were kept eligible for the

The Bonanza champion was exhibited by Laura and Don Harper, Wayne, Oklahoma. The win took a Limoges porcelain trophy and a $5,000 check. The Harper 1,290 pound Edmond sired entry had a phenomenal $2.03 retail value per day of age. The steer had a weight per day of age of 2.81 pounds, 1.36 yield grade and sported an 18.4 square inch ribeye.

contest with a $10 fee per head paid in March.

The final contest was held in Denver, Colorado, on July 26-28, 1973, in conjunction with the NALF annual convention. Eighty entries from 29 exhibitors in nine states divided the record purse of $14,604.00. The grand and reserve champions were presented checks for $5,000.00 and $2,500.00 respectively, along with their "one-of-a-kind" Limoges porcelain vase trophies.

Results of the show made national livestock press. Dean Jacobs best described the excellent results in his *Journal* editorial in September. "It was not a simple carcass show—merely identifying the ideal carcass without regard to performance. Nor was it an all-out performance race without consideration of composition. The judging rules merely set out to prove that steer A was better than steer B, because steer A had more retail value per day of age—a point that can hardly be argued. The 1973 "Bonanza" made it clear that the most valuable steer in the feedlot is the fastest growing, high cutability animal, and these factors on that day's beef market overshadowed quality grade from the profit standpoint.

It underscores lean yield per day as a valuable goal for the industry..."

The second Bonanza was held the following year in Kearney, Nebraska at the Fort Kearney Beef Producers feedlot owned by Carlton Noyes. The steers were slaughtered and the carcasses judged and displayed at nearby Minden Beef Company. The event drew a crowd of over 500 people.

The concept of the contest was changed from Bonanza to highlight sire groups. The feedlot performance and carcass data was fed into the Limousin sire summary data bank.

A pen of five Eclair sired steers out of Hereford cows bred by Kirby Briggs, Midland, South Dakota won the contest and brought them a check for $2,000 and the coveted grand champion trophy. (A Limoges porcelain vase.) The winning group sported a lean yield per day of age of .885 and 51.32 percent cutability. In the feedlot they gained 2.96 pounds per day and logged a weight per day of age of 2.66.

The feedlot gain trophy went to Symens Brothers, Amherst, South Dakota on their pen of Diplomate-Red Angus crosses. They had an average daily gain

Dean Jacobs, International Limousin Journal *staff, who programmed the competitive features of the innovative Bonanza contest making placings on retail value per day of age. He is pictured with Steve Garst, Garst Farms, Coon Rapids, Iowa. Steve served on the NALF board and purchased the Bonanza champion, donating the cuts to the NALF office staff.*

Five Eclair sired steers out of Hereford cows bred by Kirby Briggs, Midland, South Dakota won Bonanza II in 1973. The contest was designed to measure sire groups in gainability and other economically important traits for profit. Kirby and Lillian Briggs are pictured with the champions.

of 3.43 pounds and a 2.70 weight per day of age.

The judging called special attention to one individual carcass in the contest. A steer entered by Don Bohr, Wessington, South Dakota in his group of Dandy x Shorthorn steers produced an 842 pound carcass with an 18.9 square inch ribeye; .4 inch fat cover; yield grade of 1.1; quality grade choice and a lean yield per day of age of 1.022 pounds. The steer was 447 days old at time of slaughter and had a cutability percentage of 54.26.

The contest attracted widespread attention and did much to bring the word Limousin into the minds and on the tongues of the industry. The industry was taking this breed more seriously because of two Bonanza events.

By this time the carcass value of Limousin was making strong inroads into the steer shows. Many breeders were having successful club calf sales. Bud Prosser, Mead, Colorado, was chairman of the NALF steer and carcass show committee. He proposed putting a $1,000 premium for Limousin champions at approved major shows; the board approved. It eventually got expensive because the breed started having a lot of champions.

Prosser was a breeder himself; he had been to France and knew Limousin cattle. He was the driving force that encouraged Bov Import, his employer, to enter steers in the Great Western Expo. He sorted the steers to go in the test.

It was Prosser's efforts however, conducting seminars over the country, that played such an important part in breed growth. He combined lectures and demonstrations on improving A.I. techniques on the ranch and selling semen—Limousin semen. He spent weeks in South Dakota. Jim Easland, DeSmet, South Dakota would set up the meetings in banks, sale barns or wherever and Bud would come in and take the stage. The next day they would visit ranchers across the country until time to take the podium for the next meeting. Many of the breed's successes can be attributed to Prosser's efforts and understanding.

Jim Easland might not have invented the contract herd concept, but he certainly refined the system. Once he found a good herd of commercial cows he either sold them Limousin semen or he furnished the semen and made a contract on the resulting calves.

The National Western Stock Show is held in Denver, Colorado the early part of January each year. If it is not the world's largest cattle show it can certainly be said it is the most complete when the show in the yards of carloads and pens of bulls,

Limousin cattle were displayed in front of the Livestock Exchange Building in the yards at the National Western. The 20 acre yards display of carloads and pens of bulls, commercial feeders and commercial heifers makes the show unique. The National Western complex encompasses 57 acres.

feeder cattle and carcass entries are combined with the show of purebreds and show steers on the hill.

The first time I attended the show there were over 100 carloads of purebred bulls, many loads of feeder steers and fat cattle in addition to 28,000 head of cattle in the yards for the market that day.

The National Western was founded in 1906 and has been held each year without a break, even spanning both world wars. It has been a cattleman's show from the beginning. The NALF office is located in the old Exchange Office of the yards. It is no surprise that great effort was made to make a favorable breed impression at this yearly event.

The first National Limousin sale was held in the yards in 1971 and has been a kick-off event of the show each year since then. Most years it has been the highest averaging breed sale of the show.

At the 1972 event, Jerry Adamson had a load of Limousin-Angus calves bred at his Rocking J Ranch, Cody, Nebraska. This load (15 head) captured the imagination of cattlemen and when auctioned they sold for 63 cents per pound more than the grand champions of the feeder show.

Lee Leachman, Hotchkiss, Colorado brought a load of calves to the show the following year. They were picked grand champions of the show and set a new National Western record when they sold for $1.59 per pound. On the hill that same year a Limousin-Angus steer was champion and reserve grand champion of the show.

Adamson was not through with supplying the breed with promotion material though. In 1973 Rocking J feeders captured the grand champion pen of five feeder steers award at the three major livestock shows in the United States: The Eastern National Livestock Show, Timonium, Maryland;

Lee Leachman, Hotchkiss, Colorado exhibited the 1973 grand champion carload of feeders at the National Western Stock Show. The Dandy x Angus load set a new show record when they sold for $1.59 per pound.

The American Royal Show, Kansas City, Missouri; the Grand National Exposition, San Francisco, California. The calves were all sired by Diplomate (X Alaska bred by Bourbon). His grand champion pen at the National Western the following January were also by Diplomate.

Dameron Land and Livestock, Salida, Colorado showed the first grand champion carload of bulls in Denver. In the first four years Dameron exhibited the grand champion carload three times and had the reserve carload the following two years. The 800 head Dameron cow herd was dispersed in 1979 for 1.03 million dollars.

In the summer of 1973, Bob Vantrease made a visit to the National Western management office and discussed the possibility of moving our breeding cattle "to the hill" (in the coliseum instead of in the yards) with the rest of the purebred breeds for the 1974 show. "It seemed that we needed a lot more exposure for our breeding cattle," Vantrease says. "They informed me they would decide when the Limousin breed was going to make a contribution to the cattle industry and when they did they would allow us to get into the show in the coliseum. Needless to say, I was not very happy with their decision.

"On the way back to my office I ran into Marvin Hine who was the owner of the Livestock Exchange Building which was right next to the National Western Stock Show at the entrance to the yard. He indicated they were thinking about leasing out the parking lot in front of the building for exhibitors of cattle. When I heard that I just rented the whole area for the coming year's show. We set up portable panels in the lot and that next stock show we were highly visible as a breed because we had the best possible location there. Every cattleman that came down the hill either into the yards or into the Ex-

change building had to pass right through our exhibits.

"The following year," Vantrease continued, "we added both the carload of Limousin bulls and pens of Limousin bulls to the show in the yards. Limousin were the only breed making that exhibit other than the traditional English breeds. Bill Dameron, Salida, Colorado had the grand champion carload on 75 percent Limousin bulls with adjusted weaning weights averaging 600 pounds. They were sired by Dandin C, Dandy and Espoir CIM 8. They all sold in the yards at private treaty before the show ended.

"The champion pen of three bulls was bred by Yackley Limousin, Onida, South Dakota. It was the first of a long string of Limousin bulls wearing the YK brand to win a championship in the coveted yard event."

During this period of our breed progress, Limousin steers were winning a big share of the steer shows around the country. One account of these successes was included in the NALF newsletter of the March 1974 *Limousin Journal*.

"Steer Shows Fantastic! In the winter of 1974 we had a fantastic run of champions. Phoenix—Champion steer on-foot—sold for a record price. Fort Worth—Grand Champion steer on foot—sold for a record price. Tampa—Grand Champion steer on foot. San Antonio—Grand and Reserve Grand Champion on foot and grand champion carcass."

The inter-breed competition successes provided the base for much of the breed promotion handled through hand-out brochures, a Limousin movie, booths for fairs or meetings and individual breeder promotions.

Our breeding cattle shows were beginning to take on more importance. Numbers and quality were increasing and the cattle were being fitted more professionally. Our first breed shows were an exercise in futility. Many were the times when Booth, Holloway, Vantrease and I would take off after the show wondering how long it was going to be before we would see our classes with all solid red cattle in them. How long before all the horned cattle would show up dehorned? When would the white faces be bred off?

Early in the breed history the board voted to show purebreds (87 percent females, 93 percent bulls) in the same classes as the pure French cattle (fullbloods). Both the full French pedigrees and the bred up American purebreds would carry "purebred" on their pedigree. It was not long until breeders with full French bred cattle were asking the *Journal* to list their entries in their advertising as "fullblood." They requested the same designation for their cattle in sale catalogs. Today it is common practice to use the purebred and fullblood designations even though many of our purebreds are approaching the 100 percent in French blood. Percentage animals had their own classes and champions. When the purebreds started to show, we saw great improvement in the pattern of the cattle in the purebred show. They were sounder on their feet and legs and more level in their rumps.

A major difference in our breeding shows however was the influence of the Bov sire Eclair (X Castor bred by Chastanet) whose get started winning heavily. At the 1974 World Futurity, Eclair get dominated the show. His get won seven championships and placed first in ten of the twenty-six classes. He sired both the grand champion percentage bull and female. He sired the first and second prize get.

In the 1975 National Limousin Sire Summary Eclair posted the highest yearling weight, cutability and retail cuts per day of age ratios. His heavy use as a sire in the breed made his influence of even greater significance.

"It was evident that more and more breeders were demanding sire information—ie—birth weights, calving ease, weaning and yearling weights, feedlot and carcass information," Vantrease relates. "We went to work to acquire the data. By this time it became evident we were going to require a technician (scientist). I called Dr. Gary Minish at Virginia Polytechnic Institute, Blacksburg, Virginia and he recommended a student who was just receiving his PhD in animal breeding. Dr. Larry Benyshek was hired to collect the data and to produce a sire summary for the NALF. The Simmental breed had just published the first sire summary and Dr. Benyshek produced the second (first for Limousin). He is now recognized as one of the leaders in the field. He is currently on the staff at the University of Georgia. His abilities and access to the NALF records certainly led the way for much research in the field of sire evaluation. His program developed over the

Limousin steers were grand champion at Illinois, Iowa and the Ozark Empire, all major Corn Belt shows, in 1973 and 1974. Pictured with the 1973 Ozark Empire grand over all breeds is Keith Kissee, Ozark, Missouri.

Eclair CIM 51, by Castor out of Turquoise by Mignon, was bred by Chastanet and imported by Bov Import. The 1975 Sire Summary listed his impressive performance: birth weight 97.5; 205 day weight 103.2 (third highest); 365 day weight 103.1 (highest); cutability ratio 101.5 (highest); pounds retail day age ratio 101.2 (highest). For several years his get dominated the U.S. show rings and auction sales. He compiled records that read "...sons and daughters won 10 of 26 classes; ...took 7 champion ribbons at the 1975 Limousin Futurity."

years is now used by several other breeds in the United States.

"After performance data became optional for registration, it was imperative to encourage the use of records as a tool for selection. The program needed a name. Simmental had SMILE, Charolais had CHIP, etc. We wanted to find an appropriate name. We thought the word BEEF was appropriate. We spent hours and days toying with matching words to the letters B E E F; hence we decided on Breeding Efficiently with Essential Facts.

"The use of this program (BEEF), the hiring of Dr. Benyshek and the publishing of the sire summary began to create some recognition of Limousin among the academia. However, there was more work to be done.

"Two universities that were very influential in our industry were Oklahoma State and Texas A & M. We decided to go to work with them. A test was designed to compare Limousin to the popular black Baldie (Hereford X Angus cross). We approached Texas A & M to handle this project. The 36 head of Limousin cross steers were all donated by breeders; they were a very average set of steers. The black Baldies came from the famous Pitchfork Ranch and were hand picked by the manager Jim Humphreys. The test was conducted at the Spur Experiment Station. The results were conclusive. The Limousin steers outdid the others in every category except quality grade. They couldn't believe it, but they had the evidence. However, it was hard to get them to publish any information even after the test was final. Reluctantly, they did.

"Now that we were beginning to substantiate what these big reds could do," Vantrease said, "it became apparent we needed to compile all of this data and produce it in a text form. The board approved a $20,000 grant and awarded it to Oklahoma State University, Stillwater, Oklahoma.

"Dr. Richard Frahm, one of the leading geneticists in the U.S., took charge of the project. It wasn't long until he was impressed with the data and be-

Carcass Value, Feed Costs and Return Per Steer Above Feed Costs in the Texas A&M Test.

Carcass Price Set	Slaughter Group 1			Slaughter Group 2		
	¾ Li	½ Li	A-H	¾ Li	½ Li	A-H
$0/cwt differential for yield grade						
Carcass value	526.68	490.65	506.54	566.65	541.59	538.93
Feed Cost	241.20	224.19	225.63	280.74	283.72	253.76
Return per steer	285.48	266.46	280.91	285.91	257.87	285.17
$2/cwt differential for yield grade						
Carcass value	549.16	514.50	514.79	492.00	564.44	546.48
Feed Cost	241.20	224.19	225.63	280.74	283.72	253.76
Return per steer	307.96	290.31	289.16	211.26	280.72	292.72
$5/cwt differential for yield grade						
Carcass Value	582.88	550.27	527.18	633.13	600.21	557.82
Feed Cost	241.20	224.19	225.63	280.74	283.72	253.76
Return per steer	341.68	326.08	301.55	352.39	316.49	304.06

Reprint *International Limousin Journal* July 1977.

Notes:
1. Results showed both groups of Limousin steers continued to gain at the same rate after the first slaughter group was taken out at 177 days. The Angus x Hereford steers increased less in total weight gain to the 212 day second slaughter.

2. In general terms Limousin groups gained one-half pound per day more than the Angus x Hereford steers and were 20 to 25 percent more efficient.

3. The market at the time of the test conclusion had a $6.00 spread between yield grades and $1.00 spread between quality grades. The test clearly shows $40.00 more per head in favor of Limousin.—notes by Editor

Carcass Weights and Yield Grade Factors According to Breed Groups in the Texas A&M Test

Characteristic	Breed Groups		
	¾ Li	½ Li	A-H
Slaughter wt (lb)	1082.8	1003.3	1028.6
Warm Carcass wt (lb)	694.6	635.5	650.9
Chilled Carcass wt (lb)	680.8	622.8	637.8
Dressing Percentage	62.9	62.1	62.0
Yield Grade Factors			
Fat thickness (in)	0.23	0.29	0.54
Adj. Fat thickness (in)	0.15	0.27	0.55
Ribeye Area (sq. in)	16.1	14.1	12.4
Est. % KPH	2.2	2.8	3.2
USDA Yield Grade	0.9	1.6	3.0
Est. % Boneless Cuts	54.8	53.3	50.0

Reprint *International Limousin Journal* July 1977.

Notes:
1. Cutability tests showed Limousin to be much more desirable and produced a higher percentage cut-out from the high priced cuts.
2. Marbling scores and USDA quality grades were significantly in favor of the Angus x Hereford steers.
3. Taste panel tests indicated all groups were acceptable in eating quality, even those Limousin steers that did not grade choice. This reinforced other studies done prior to the A&M test.—notes by Editor

Hereford, Angus and Hereford x Angus cows were bred to bulls from eight different beef breeds. The study considered total production costs and included calving ease, calf liveability, growth rate, feed efficiency, carcass composition and quality grade on over 2,400 matings to compute the cow-profit study. Limousin sired calves had from 16 percent to 100 percent less calving problems than Blonde d'Aquitaine, South Devon, Simmental, Charolais, Maine Anjou and Chianina bulls tested...data from Oklahoma State University Research Bulletin B736, February 1978. Copy available on request.

came extremely involved with the breed. Dr. Frahm went to France and came home a solid believer in Limousin after seeing the herds and reviewing a substantial amount of Limousin performance data compiled by French researchers. Under his direction we were able to start a progeny test for about ten young sires each year on the OSU research herd.''

Dr. Frahm relates to us the story of his early research in the U.S.:

"First, the history of my involvement in doing a research study in 1976 and 1977, that culminated in Oklahoma Agricultural Extension Station Bulletin B-736: *An Evaluation of Limousin Cattle,* that was published in 1978. In the fall of 1975 Bob Vantrease, the executive vice president of the North American Limousin Foundation, contacted me and asked me to review research with Limousin worldwide. He wanted me to summarize and interpret this, relative to evaluating the role of Limousin cattle for beef production in North America. I was a bit hesitant about getting involved in such a project at that time for several reasons. Among them were: 1-I was already heavily committed to research of my own which was in progress; 2-I knew there was little experimental data involving Limousin cattle available in the U.S., and had no idea whether there was any meaningful data in other countries; and 3-some skepticism as to how such a summary would be used by NALF. I, like most geneticists, was not interested in having selected portion of data be used for breed promotional activities. Bob had obviously given this much thought and persisted in discussing the feasibility of such a project.

"Vantrease convinced me that the board members of NALF were interested in an unbiased, scientific appraisal based on available experimental data. He was convinced that data existed in several countries and thought I should visit these countries to talk with the researchers involved, as well as seeing, firsthand, Limousin herds in France. Thus, based on these discussions, I wrote a proposal for a two-year research project that was approved by NALF for $20,000. This money was spent for: 1-apprenticeship for two years for a M.S. student to assist me (Danny R. Belcher); 2-my travel to collect pertinent research data; and 3-publication of the bulletin. During the spring of 1976 I visited NALF headquarters in Denver, and also as many Limousin breeders as possible

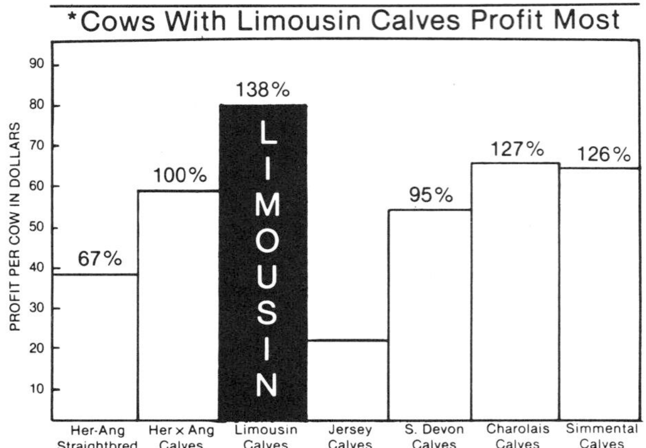

Consider the Limousin case for calving ease:

Incidence of Calving Difficulty

LIMOUSIN —	16.7%	less than Blonde d'Aquitaine
LIMOUSIN —	31.3%	less than South Devon
LIMOUSIN —	40.6%	less than Simmental
LIMOUSIN —	88.7%	less than Charolais
LIMOUSIN —	96.5%	less than Maine-Anjou
LIMOUSIN —	100.0%	less than Chianina

"averaged over seven studies including 29,789 calves, the Limousin-sired calves had less calving difficulty."

— OSU Research Bulletin B736, February 1978.*

The profit per cow chart above was used widely in NALF promotion. The data was collected in the Frahm Oklahoma research. The profit per cow on progeny performance for a post weaning feeding period of 217 days and $1.15 per pound for choice grade and $1.07 per pound for good carcass retail product.

to learn as much as I could about the Limousin breed as it was developing in the U.S. Bill Dameron, of the Dameron Land and Cattle Company, Salida, Colorado, was particularly helpful. Bill was kind enough to provide us with an extensive set of pelvic measurement data that he had collected on Limousin cross heifers during his grading-up program. Danny Belcher analyzed this for his M.S. thesis in 1979. *(Factors affecting calving difficulty and the influence of pelvic size on calving difficulty in percentage Limousin heifers.)*

"During the summer of 1976 I spent three weeks mostly in France, but also in Italy and England, visiting Limousin breeders and research stations with experimental data on Limousin. It was a real pleasure to spend considerable time with the then-president of the International Limousin Association, Louis de Neuville. That summer I also visited

research stations in Canada that had considerable research data on Limousin. Most of 1977 was used to study, analyze, summarize and interpret the data that had been gathered.

"I was invited to attend the International Limousin Conference in Limoges in September of 1977, where I presented a summary of my research. The rest of the fall was spent completing the project and writing the bulletin. I made a verbal presentation and delivered a typewritten copy of the final report to the NALF Board of Directors in January of 1978. The bulletin was published later that spring. In fact, the first copies of the bulletin were presented to the breeders that attended the Oklahoma Limousin Breeders Annual Banquet in March of 1978.

"You asked me if I was pleased with the results of this study. I can give an unqualified 'Yes' to this," Frahm stated. "The cooperation of the Limousin people that helped me get the necessary experimental data was outstanding. I made many good friends while doing this study. The most pleasing thing about this study was the positive way in which the NALF has used the study. The bulletin has been widely distributed. NALF has re-ordered printings several times (10,000 at a time). I have had more requests for copies of this bulletin than any other research paper I have written. I was so impressed with the Limousin breed that I have used Limousin bulls since then in my research at OSU.

Animal scientist Richard Frahm, Oklahoma State University, headed a NALF project of "collecting and analyzing all available experimental and performance data on Limousin in the world." Printings of his work exceeded 10,000 copies. Dr. Frahm was recently appointed chairman of the Animal Science Department, Virginia Polytechnic Institute, Blacksburg, Virginia.

"You asked about my evaluation of the Limousin breed, based on the data available to me. I gave an overall evaluation of the Limousin for various traits of economic importance. For the most part, data that I and others have collected since then generally tend to support that evaluation. The Limousin breed has certainly been accepted by the U.S. cattle industry and has many attributes that make it a logical choice in many planned crossbreeding systems. Its real strength, of course, is for use as a sire in a terminal crossbreeding system. But there may well be many production situations where Limousin could be successfully used in a rotational crossbreeding system. The Limousin breed has been fortunate to have so many breeders oriented toward genetically improving the breed for economically important traits. I am pleased that many are following my recommendation to select for improved growth rate, while being careful to avoid increasing weights when possible. Smaller birth weights and correspondingly less calving problems are distinct advantages of Limousin bulls, and care must be taken to preserve these traits. With sounder breeding selection programs, the Limousin breed will make a major contribution to beef production in the U.S. for a long time to come," Frahm concludes.

Vantrease says, "The Frahm study gave the Limousin breed much credibility among the university people who doubted our value to the industry up to this time." The studies that were done by Dr. Frahm for NALF include: Comparison of Charolais and Limousin as Terminal Cross Sire Breeds. Birth, weaning, feedlot and carcass traits were evaluated on 1,181 calves sired by Charolais and Limousin bulls out of eight crossbred dam groups of Hereford, Angus, Simmental, Brown Swiss and Jersey. The calves were born over a four-year period. The results were: Charolais were heavier at birth, had a 9.9 percent higher incidence of difficult calvings and a 4.6 percent greater preweaning death loss than the Limousin calves.

The Charolais calves gained 31 grams per day more and were heavier at weaning. Feed efficiency was equivalent on a grade equivalent basis, Charolais crosses produced 7 kg. heavier carcasses and had 22 grams more carcass weight per day of age. (J.M. Dhuyvetter, R.R. Frahm)

"The report we put together was a cohesive report," Danny Belcher, OSU graduate student who compiled the Limousin data, says. "The comparisons we made 14 years ago are not relevent now with so many new bloodlines. The research we did was performed on smaller-framed cattle with more muscle distinction, muscle volume and problems with structural correctness. Our research was directed toward the commercial cattleman."

Dr. Larry Benyshek designed and programmed the first NALF Sire Summary and has up-dated the program each year since the first one in 1973. His EPD program produced with the new "Reduced Animal Model" was first with the Limousin Sire Summary in 1985 and has since been adopted by several breeds.

Chapter 3
Fuel for the Limousin Fire

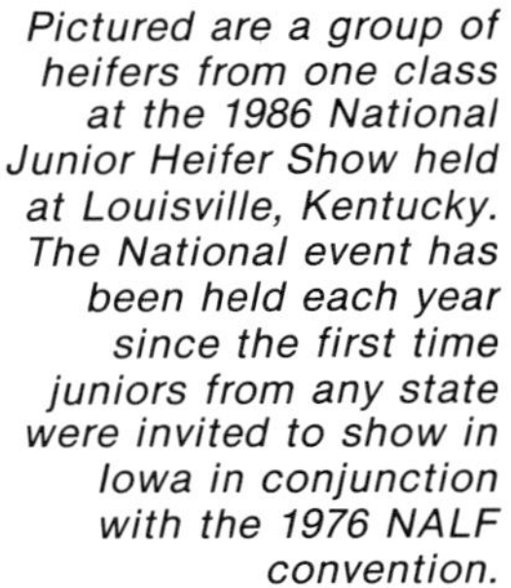

Pictured are a group of heifers from one class at the 1986 National Junior Heifer Show held at Louisville, Kentucky. The National event has been held each year since the first time juniors from any state were invited to show in Iowa in conjunction with the 1976 NALF convention.

A colored preacher in Virginia once told me, "When the ladies of the church quit having church suppers to buy new song books and the men aren't having meetings to raise money to repair the roof, you have a dead congregation."

Involvement was the key to his successful parish. Bob Vantrease was a staunch supporter of this philosophy and practiced it faithfully. In his newsletter each month, he was constantly encouraging the readers to join the NALF, be active in their respective state associations, consign good cattle to their state sale, or attend the national conventions. He talked to the women about becoming active in the ladies auxiliary and to the young people about getting a Limousin heifer and joining the junior association. His primary job was to administer the herd book but he counted heavily on the people to make the business grow.

The parent of Limousin people activity was the North American Limousin Foundation. The national convention was its major homecoming. In Tulsa in 1974 the convention was longer than others because the First World Limousin Futurity was added to the events. The successful 82 head event averaged $3,812. Because of the longer convention, more social time was added to the activities. Country western singer and entertainer of the year award winner Mel Tillis entertained at the banquet and dance. The NALF conventions had now become a family affair.

It was at the 1974 convention following a most successful ladies luncheon that the ladies Limousin auxiliary, later to be known as Limouselles, was founded. Bonnie Booth, Welch, Oklahoma was elected its first president.

The Oklahoma Limousin Association, pleased with the Tulsa results, were given the opportunity to host the 1975 convention in Oklahoma City. The second International Limousin Council met during the convention and Limousin visitors and boosters from 10 countries attended the convention before making an extended visit of farms, feedlots and packing plants here in the United States. Their presence and participation added interest to the annual affair.

At the 1976 convention in Des Moines, Iowa, the Iowa Limousin Association, with the cooperation of the Iowa State University Meats Department,

Limousin Prove Worth From The Inside Out!

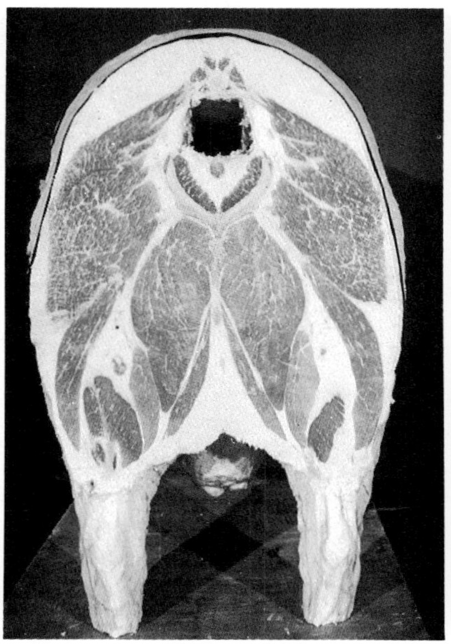

Black Steer Through The Round

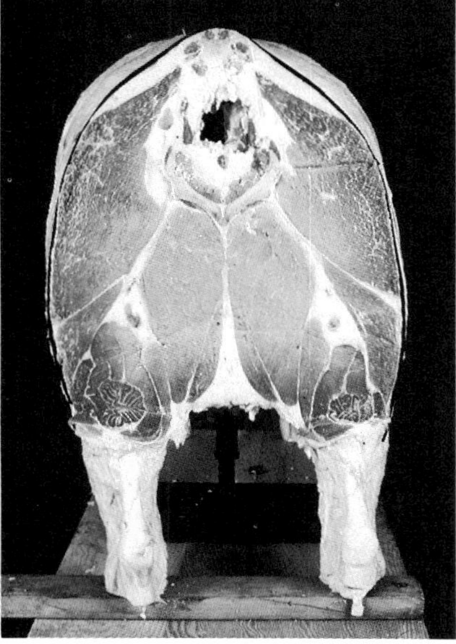

Limousin Through The Round

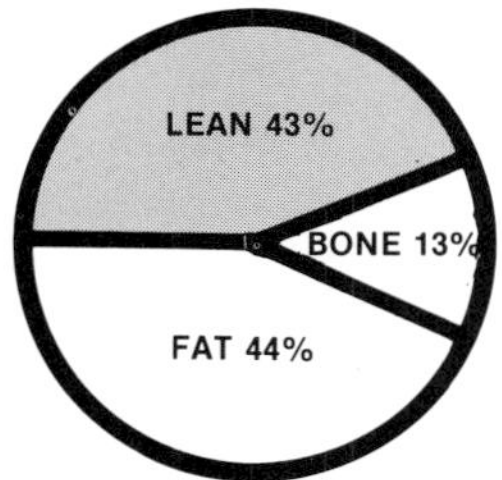

The Revealing Percentages

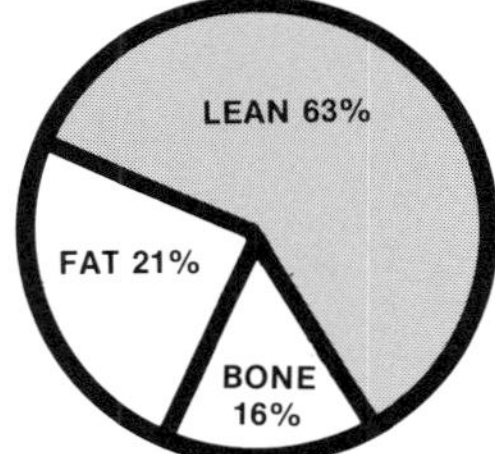

A Great Set of Figures

** A reprint from the September 1976* International Limousin Journal. *The illustrated story relates the background of this important piece of research that has graphically illustrated the real value of the Limousin bull in not only adding valued edible red meat to the carcass but genetically trimming the carcass at the same time. The illustrated data has been one of the breed's most useful promotional tools.*

One of the most revealing and interesting presentations made at the eighth annual NALF convention in Des Moines came from Dr. Ed Kline of Iowa State University. Doctor Kline and his staff at Iowa State have developed a unique way of looking at beef animals, providing a more accurate view of how live animals and carcasses compare.

In the Kline procedure, animals are slaughtered, then frozen in a standing position. They are then removed of head and hide and shanks leaving the frozen carcass in a standing position. From that point the animals are cut into cross sections at right angles to the body. Cuts are made through the bulge of round, through the stifle point, at the point of the hip bone, between the 12th and 13th rib, the fore flank (about the sixth rib) and the point of the shoulder.

Kline began his talk and slide presentation by discussing the old school which taught cattlemen to look for blocky, compact, short-legged, deep-bodied beef animals. To illustrate his point he used the slide of the black steer which appears on this page. He then progressed to the more desirable steer of today and compared a three-quarter Limousin steer with the black steer which began the program.

The three photos of each steer and accompanying charts on lean to fat to bone ratio vividly depict the differences in lean meat productivity of the Limousin carcass animal compared to the steer of yesterday. The color photo comparing the steers through the round is especially striking.

Both the animals in the comparison had a USDA quality grade of choice while the Limousin had a yield grade of two and the black steer was a yield grade 5.

For those who think we haven't come quite a way in this business during recent years, the photos on this page are irrefutable evidence that change and dramatic improvement in meat production has certainly occurred and continues to take place.

As Doctor Kline emphasized in his talk... it takes considerably less feed energy to convert to a pound of lean than to a pound of fat.

Think about the dollars the luxury of fat waste has cost everyone in the beef production chain. Interesting isn't it?

presented an "Inside Look" at finished steers. Meat scientist Dr. Ed Kline, with the assistance of several Iowa breeders, slaughtered a Limousin steer and froze the carcass in a standing position. They next made three cross section cuts of the carcass. The slide presentation comparing the Limousin sections with corresponding ones from an over-finished Angus carcass gave many convention goers their first real appreciation of the tremendous carcass value of Limousin.

The slides were incorporated in a hand-out brochure put together by Vantrease. It has proven, even to this day, to be one of the breed's most useful and popular promotion pieces. The basis for the material was made possible by members (people) who had an idea and the will to pursue it.

The Iowa State convention featured a Junior heifer show. In essence it was a national show as they invited open entries from any state. They had 40 head exhibited by juniors from eight different states.

State and regional associations are the lifeblood of the North American Limousin Foundation. They are the organizations that serve the needs of the breeders in the community, state or area through state sponsored sales, shows, field days, Limouselle functions and junior activities. They represent the Limousin breed at all-breeds cattlemen events in the area. Their effectiveness in breed growth has been substantial.

The high concentration of breeders in South Dakota, Oklahoma, Texas, Colorado and other states naturally led to organizing associations in these states early in our breed history. A sterling example of an association's effectiveness is the Southeastern Limousin Association.

Bo Worthington and Rhodes Frost at Quitman, Georgia pioneered the southeast area with the first sale in December, 1972. The Southeastern Association followed in March with a sale in Ocala, Florida. Interest in Limousin in this area was excellent because of the exposure with these two sales. You could see the area where buyers came from widen with each succeeding year.

Bob Vantrease had traveled the southeastern area of the United States before coming to work for the Limousin foundation. He says, "It was evident to me that we needed to expand the breed into other areas besides the cornbelt and the plains states. The southeast seemed to be a natural place for growth. About that time, John Spivey, McDonough, Georgia was getting into the business. I had just hired a young man by the name of Tom Gaskell. I told him to get in his car and head for the southeast and help these breeders promote the Limousin breed. Obviously it has been one of the major growth areas of the country and our extra efforts have paid rich rewards for the breed."

In the summer of 1971 the Oxandaburus of Huron, South Dakota had a field day on their ranch. They left nothing undone to make it a success. They aptly named the event Oxandaburu's Limousin Show and Tell Field Day. "We wanted to show the crowd our Limousin and tell them about the Limousin breed of cattle," Cadet said.

A crowd of over 600 people came. Special wagons borrowed from the state fair association that enabled folks to get off and on easily and shaded them from the sun were pulled by tractors to haul the crowd around the ranch. The always full wagons were taken through the feed yards and out into the pastures to see the cows and big Limousin halfblood calves. Cadet rode the tractor all the time with a loud speaker system to tell the folks all about Limousin. After a hearty lunch of Limousin hamburgers complete with all the trimmings, the crowd gathered under the shade trees to enjoy the program and the speakers. Cadet opened up the program to field questions from the audience. There were many in the crowd who were seeing their first Limousin, getting their first taste of Limousin beef and were enjoying their first exposure to the friendliness of the people in the Limousin business.

Field days have always been a great breed booster. Most are association sponsored events held on a member farm or at a fair or exhibition grounds. Some have been individual ranch events like the Show and Tell. They have always been a good place for the novice to come and learn.

Since juniors' projects have reached workable numbers, most state associations have held their state Junior Limousin Show in conjunction with their field days. Judging contests inviting FFA chapters and 4-H teams to participate have helped increase attendance. Fitting and showing demonstrations gave the novice a chance to learn. Lectures and demonstrations on artificial insemination were popular programs during the early years. The leading subject for programs changed to embryo transplants as interest in the process increased and new techniques were adopted.

The North American Limousin Junior Association is one of the very bright spots in breed growth. NALJA activities are managed by their own members. It is largely financed by contributions from individual members of the foundation. Although no figures are available for comparison, it stands to be one of the most highly endowed youth organizations of its kind in the country. The juniors also have fund raising projects of their own to put money in the till.

When the Limouselles got together in 1976 the professional models wearing the stylish clothing from the house of Lilli Rubin highlighted the affair. More than 200 ladies from several breed associations were in attendance.

The junior organization certainly started from humble beginnings. I can remember Marilyn Rinehart of Highmore, South Dakota weaving her way through the convention crowd in Denver, spotting likely prospects and pulling them out to take with her to the first organizational meeting. Gloria Jennings, Highmore, South Dakota was the first NALJA president.

There are numerous family herds in the United States today that stem from an interest in starting the children in a junior Limousin project. Many of those single heifer projects led juniors to be serious Limousin breeders in their own right when they reached adulthood. Others have responsible positions in some part of the cattle business from their experience gained in NALJA. One such junior, Kevin Oschner, Fort Collins, Colorado, recently bred and exhibited the grand champion bull at the Colorado State Fair. The bull was a great grandson of a 50 percent cow he purchased as an eight-year-old junior in 1976. Another young man from Holstein, Iowa, Mark Leonard, and who served two terms as NALJA president, is a successful breeder of Limousin today in his home state.

The most noteworthy junior activity each year is the National Junior Heifer Show. The first national show was held in Springfield, Missouri in August 1977. Steve Yackley, Onida, South Dakota, exhibited the Grand Champion. The 60 head show was a good start. The 1986 show in Louisville, Kentucky had 194 head from 19 states lead out in the competition.

The juniors put kindling to the Limousin fire in Louisville as they have done many times within their own families and in their states.

In the early 1970's our breed was getting only token recognition from the academia. Many individuals on our university staffs without any substantial data even encouraged the myth of "double muscling" fostered by supporters of the English breeds and certain of the other new breeds. It was important that our breed elicit the support and involvement of this influential segment of the industry.

Many of us who were involved with breed promotion on a day to day basis reasoned that it was our herd bulls the critics did not like. They objected to their pattern and their structure—particularly their feet and legs. We felt this was a just criticism when

Christina Baumann Massie, pictured visiting with Dr. Gary Minish after a walk through the Nordic herd. The Nordic herd was the largest pure French herd in North America at the time. Its influence on the success of the Limousin breed was substantial.

the judgement was based solely on visual appraisal of the sires that were now in Canada. With several calf crops on the ground however, we could see that many of these bulls were siring a very high percentage of acceptable offspring with regard to pattern and structure. Our successes in multi-breed steer shows was definite proof of this appraisal.

I remember talking about these questions and problems with Bud Prosser and Bob Vantrease. Bud attributed the first generation improvement in pattern and structure to the many generations of great cows behind these sires. "Those cows in France are fantastic," he said. "Every one of our critics would have to admit the mother cows in France are our kind of cows." From our discussion we concluded that 1-we could change the structure of our cattle easier than some of the other new breeds could increase cutability, gain efficiency and calving ease. Each was a heritable important economic trait that was bred into our breed for many generations; 2-we resolved to pursue the issue, each in our own way.

Vantrease made it a priority to select highly visible judges from the university ranks to officiate at our shows. "I remember one particular person, Dr. Gary Minish," says Vantrease. "He was a long time friend of mine but he didn't have much good to say about Limousin. I asked him to take a trip with me to Canada. We spent a week visiting the bull studs and Canadian Limousin herds. One of the most impressive days was spent with the Nordic herd near Guelph, Ontario.

"On the way back to Denver I could see that the trip had not created the impact I had wanted. I told him I had one more place to take him.

"The next day we drove to Bill Dameron's ranch at Salida, Colorado. After a day at this well managed breeding establishment, Gary told me on the way back, 'Now I understand why you get so excited about these cattle. I can't believe those fullblood bulls we saw this week could sire calves and yearlings like we saw today.' From that point on he was sold on our breed. Virginia Polytech has since established a herd of Limousin in their beef cattle program."

Several years before the Limouselles organization was formed, two ladies in South Dakota, Marilyn Rinehart, Highmore and Eva Oxandaburu, Huron, took a lead to get the junior and the queen programs on line. From the beginning, state associations have selected their own state queen to represent them at the national contest held during the annual meeting each year. The bright, attractive young ladies are judged on their queenly appearance and their knowledge of the Limousin breed and the business thereof. They serve as capable ambassadors of the breed in addition to adding a special flair to the pagentry during Limousin events on a state and national level.

Two events started at the National Western have

The pageant during the annual NALF convention where the National Limousin Queen is selected is a colorful event each year. State and regional contestants at the 1981 event are pictured above. (L to R) Georgia Reuer, Selby, South Dakota, Miss Congeniality; Linda Oehrke, Cole Camp, Missouri, First Runner-Up; Tina Mastropolito, Selma, California; Cheryl Linthicum (Fulkerson), Welch, Oklahoma, 1981 Queen; Jackie Hutchison, Dilley, Texas; Donna Huisman, Plainfield, Iowa; and Renee Rown, Montgomery, Minnesota.

given the breed excellent exposure. One of these has been the pre-sale party held at one of the large hotel ballrooms. The well appointed snack and hors d'oeuvre tables along with dancing to nationally recognized western entertainers was open to all who desired to enter. And the young people of all breeds did come and enjoy the Limousin hospitality. They helped carry the Limousin message in their own way and words when they told everyone about how great the people in the Limousin business were treating cattlemen of every breed.

Dee Kissler, Bennett, Colorado, took on a task of having a ladies breakfast at the Cosmopolitan Hotel in 1976 during the convention that year. She felt the women should have an event of their very own while their husbands were engaged with other aspects of the National Western show and NALF convention. She extended an open invitation to ladies of other breeds to attend. Flowers, door prizes and table gifts were collected by Dee from merchants around Denver. The highlight of the affair was a style show sponsored by world famed designer Lilli Rubin with professional models displaying fabulous high style clothing. The style show was picked up by the Limouselles and is a feature of each National Western Stock Show week.

Dan and Dee Kissler, Bennett, Colorado are working organizers. Dee served two years as Limouselle president and her style shows were classics. Dan actively organized convention and yard facilities for the annual National Western event for several years.

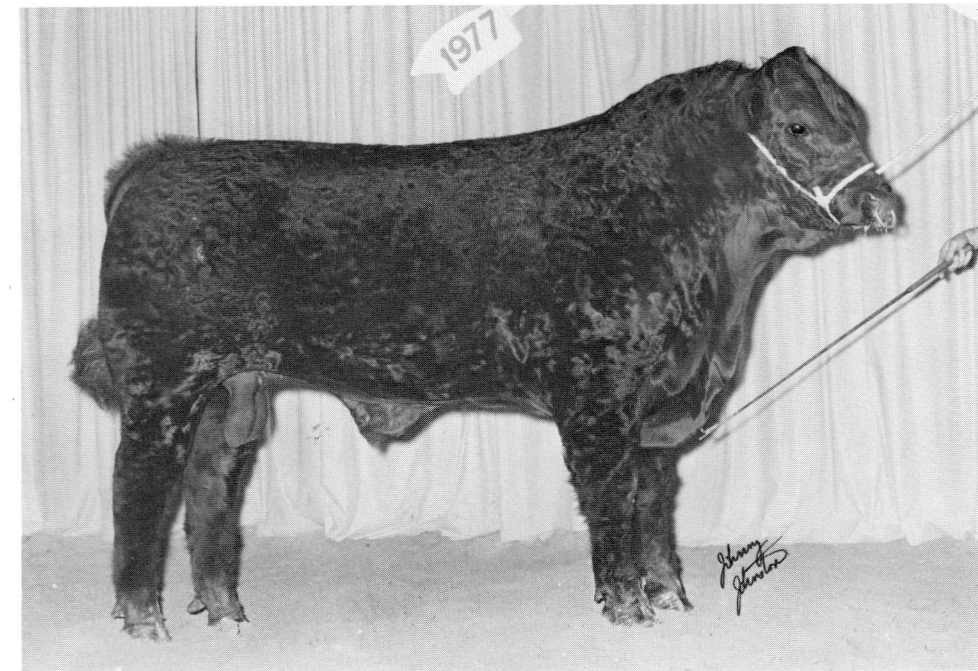

YK Black Jack attracted widespread attention when he topped the Yackley sale at $6,550 in 1976. The Eclair son out of a Dandy x Angus cow had 584 and 1,110 pound weaning and yearling weights. Holcomb and Yates, the buyers, used him successfully in their club calf breeding program near Stanton, Texas.

Chapter 4
Selling the Limousin Breed

All cattlemen were experiencing tough conditions in the mid 1970's. Despite the depressed cattle market, the demand for Limousin was still on the up-swing. Registrations had reached 29,820 and transfers issued numbered 14,109 for Limousin in 1974. Four hundred ninety-eight new memberships had been granted. The National Sale, which at times seemed to be a barometer of activity, had averaged $3,702. The first carload and pen show of Limousin bulls in the yards at the National Western had been successful. Larry Hollers, Cody, Nebraska had exhibited the pen of crossbred calves by 75 percent Limousin bulls that topped the feeder sale of the National Western.

The Iowa State Association sale only averaged $485 in 1975. That same year Yackley's sale of performance bulls averaged $513. (1976 average $1,232.) It was a good time to buy Limousin and promotion efforts were expanded lauding our breed strengths.

Alan Richardson, Clayton, Indiana, who later became a NALF board member, was one of the many cattlemen who decided to switch to Limousin during this time. "We had Shorthorn cattle and changed to Limousin for three reasons: growth, muscle and a marketable animal," Richardson recalls. "A breed change is not an easy thing to accomplish," he said. "The open artificial insemination policy in the breed was something I liked. I could use any Limousin bull in the United States or Canada. We started in 1975 when prices were depressed and that allowed us to buy good cows at a very reasonable price."

Limousin show successes around the country and the ad campaign, even though it was operated on a meager budget, was creating a great volume of mail seeking information on the breed. Bob Vantrease had created a multi-page mailing piece in 1973 and again in 1974 to answer these inquiries and use as pick-up pieces for fairs, shows and conventions.

In late 1974 the supply was exhausted and a new printing was needed. Vantrease came to the *Limousin Journal* office in Fort Collins to talk about his plan of producing a quality 4-color brochure of 20 pages with an attractive cover that would give a history of Limousin, recount some of our test station data and generally promote the Limousin breed. We

Lefty Elliott, Fresno, California (pictured with wife, Judy) served as NALF president and was also the first president of the Western States Limousin Association. His influence in expanding the breed in the western states through his Rancho McLin herd had a lasting impact.

prepared the lay-out and assisted in collecting the copy.

I was invited to the fall board meeting to discuss the production of this piece of literature. I proposed to the board that they incorporate the 24 page piece in the February '75 issue of the magazine. I told them if they did this I would collect 2000 new names of people who had shown an interest in Limousin and send these folks a copy of the February issue and two other issues of the magazine free of charge.

One of the board members asked how I intended to acquire the names. I described a plan to make a mailing to all active NALF members and ask them to submit ten names and addresses of friends, neighbors or associates who might be prospective members or buyers of Limousin cattle. I would urge them to include their banker and to be discreet in selecting the best list of prospects they could put together.

Lefty Elliott, Fresno, California, hardly let me finish my presentation before he complimented me on the plan and asked what I would charge to send the magazine to this proposed list of names for one full year. My reply was ten cents over the cost of

a half price subscription rate; the reason for this was to serve a postal requirement so the *Journal* could qualify the additional names for our mailing permit.

The discussion was brief before the board approved the Report '75 and bought 2000 subscriptions. This plan worked so well that it was renewed each year as long as I owned the *Limousin Journal*. Each year we solicited the old list for renewals and asked members for a new list.

Two years after starting the program we conducted a survey of buyers at sales that averaged in the middle of the road price range. Our survey revealed a very high percentage of the non-member buyers were already receiving the *Limousin Journal*. It was a compliment to the magazine and the far-sightedness of the NALF Board of Directors.

It was during this difficult marketing period that the board decided to produce a movie. "There were lots of activities that were keeping our breed moving during the depressed cattle markets of 1973-74-75," recalls Vantrease, "but I think the movie '7000 Years New' was one of our major accomplishments. Something was needed to fulfill the requests for information on the breed from universities, 4-H and FFA clubs, cattle meetings and whatever. This movie filled the need.

"We filmed about half the movie at Jerry Adamson's place in Nebraska and then went to Ben Johnson's in Oklahoma. Ben is a tremendous cowboy in his own right as he was the World Champion Calf Roper in the early 1950's. He is also a

M.E. "Marvin" Singleton, Waxahachie, Texas started in the Limousin business in 1969. He is recognized as the founder of the Texas Limousin Association. His importation of fullbloods Bandito and Beauriel from the Brandon Research Station were some of the early imports into the United States.

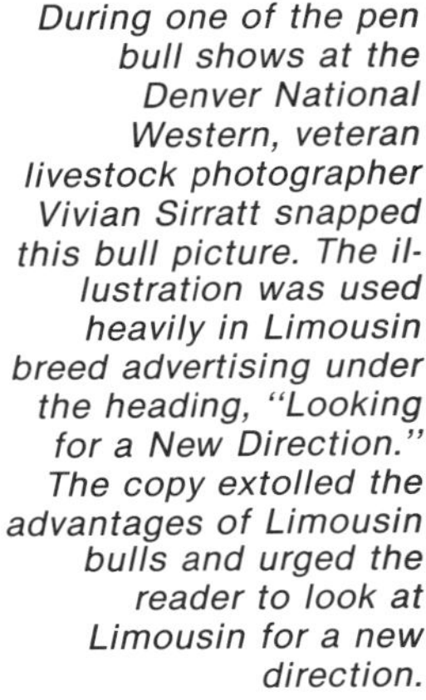
During one of the pen bull shows at the Denver National Western, veteran livestock photographer Vivian Sirratt snapped this bull picture. The illustration was used heavily in Limousin breed advertising under the heading, "Looking for a New Direction." The copy extolled the advantages of Limousin bulls and urged the reader to look at Limousin for a new direction.

movie actor of noteable fame in Hollywood circles, receiving an Oscar one year for Best Supporting Actor. He had a herd of Limousin cattle in Oklahoma and I went down and made a deal with Ben to let us shoot some film on his ranch. He also agreed to narrate the movie for us. His presence and expertise made it a class production. I think it did a tremendously important function for us in promoting our breed which was necessary at that time.

"I will never forget," Vantrease continues, "we had worked on this thing for about six months and we wanted to premier it at the NALF party at the Stock Show in Denver. The movie company in Texas that did all the filming, editing and all the other work that had to be done to get it produced right had shipped the first copy to me that afternoon. It arrived at the Denver airport at about 8:00 that night. I went to the airport to pick it up and get it back to the Cosmopolitan Hotel where the convention party was going on. We got the equipment set up and showed it to a standing-room-only crowd. When it was over they cheered and applauded and asked me to run it over again. We could hardly make enough copies of the film fast enough to get it in the hands of those folks who wanted it right now."

The breed advertising in the national livestock press started on a low budget using five weekly tabloids and one monthly magazine. All were multi-breed publications. In the early advertising programs the information was institutional in nature, using test data to substantiate our claims for the breed. "Proof not promises" was an example.

Generally the ads would run two times a year in the respective publications. A bull ad would appear before the bull sale season and a cow, feeder or feedlot ad before the calf marketing period in each particular area we were trying to reach.

As the budget increased, more publications were added to the list and more insertions alloted to the basic big five. Inquiry increased markedly when the Oklahoma Research bulletin produced by Dr. Frahm was offered as an ad premium. Most of the new publications we decided to use were also multi-breed and had a concentration of circulation in a particular state. An institutional ad would be sent to these papers to run before the annual state sale and meeting to bolster Limousin exposure of the state association ad in the same issue and to help promote the sale and breed.

Providing attractive Limousin booths to the associations and members was another project taken on by Vantrease during this period. The booths would be sent from fair to fair across the country and were used as an information headquarters for the breeders. More importantly, it served as an attraction to the Limousin cattle stalls and as a base to answer inquiries concerning the breed and pass out breed promotional material.

"It took quite a while to cover all the avenues of

breed promotion,'' Vantrease says. ''However, by this time we were coming out of that rotten cattle market in the late '70's. We had a good line of promotional material for the state associations or the individual breeders themselves to use upon request. We had the booths, slide shows, show barn banners, a movie, hand-out brochures, Limousin hats and a good selection of novelties and other items for them to draw on to further their promotional efforts.''

Chapter 5
The Fullblood Expansion

As the cattle business was coming out of a price slump in 1975, new interest was shown for pure French bloodlines. A steady trickle of fullbloods were coming out of Canada. However, the International Breeders Service, Inc. (IBSI) sale in Oklahoma City, Oklahoma whetted the appetite of fullblood buyers with an offering of 21 head of fullbloods. The sale in Oklahoma City was held in conjunction with the NALF Convention in 1975. The importation by Raymond Hefner, et.al., came from England, Sweden and Norway.

IBSI had a previous importation the year before of 31 head which sold at private treaty to Woody Sudbrink, Top Hat Ranch, Madison, Florida. This draft was the start of a sizeable herd for Sudbrink.

In 1975 Nordic Farms, Guelph, Ontario, had scheduled a fall sale. Two Texans, Ben Woodson, Del Rio and Hayes Mitchell, Marfa, went to Canada and bought the offering at private treaty. They moved 19 females and bulls to the newly established herd, Deux Amis (meaning two friends in French), near Luling. The remaining 10 foundation females of the purchase went directly from Nordic to a transplant center before coming to Texas. From this beginning many great producing cows came into U.S. herds including Nordic Debonnaire. Her many ET daughters have written new breed records in performance and in the show and sale rings of the country.

The December issue of the *Limousin Journal* in 1975 reported the fact that 160 fullblood Limousin females and 200 fullblood bulls had been recorded to date in the NALF herd book. In the National Sale and other bright light events the fullbloods of acceptable quality were selling higher than the purebreds. With the refinement of embryo transplant procedures and the advancing fullblood market, many investor type breeders entered the scene. The fullblood investment was greater but the promise for profit was greater also. The biggest percentage of buyers, however, were breeders who added a

The Deux Amis principals and their wives pose at the celebration that welcomed the arrival in Texas of their fullblood purchases from Nordic Farms, Guelph, Ontario. Pictured are Maxine and Hayes Mitchell, Marfa, Texas; Christina Baumann Massie, Nordic Farms; and Ben and Shirley Woodson, Del Rio, Texas.

fullblood program to their existing purebred enterprise. In the 1975 sale season, each of the 25 top selling females at auction in the United States and Canada were fullbloods.

The supply of purebreds was dramatically increased with the advent of embryo transfer. Merle Derochie's Porcupine Hills Transplant Center near Calgary was one of the largest operations in terms of volume.

In the Derochie program with the customer furnishing the semen the published price for the first recipient was $1,900 and each additional one from the same flush was $1,700. In another program at this center, the pregnant recipients would be divided equally with the customer. In this plan the first recipient from the flush going to the donor owner would cost $900 and each additional one $500. All recoveries during this era were done surgically. Porcupine Hills advertised in 1976 that they were having a 74% rate of conception on their recipients at 90 days.

In 1975 Mel Hatley, Lexington, Oklahoma bought 120 pregnant recipients from the Calgary firm. In 1976 and 1977 Woody Sudbrink and Fred DeMier, Miami, Oklahoma bought drafts of 140 and 110 head.

Mel Hatley put together a sale offering of embryo transplant fullbloods from their Canadian purchase.

The seats were full with buyers coming from eleven states.

The 30 pure French calves were by four different sires. Eclair sired 27, Fanfaron, Espoir de Carnaval and Essor each sired one. There were 14 different dams represented. The calves averaged 9.2 months of age.

Bob Vantrease got a syndicate of 11 breeders together who purchased MH Bold Tactic (later changed to Gibraltar) for $55,000. The 19 bull calves averaged $7,576 and the 11 heifer calves brought $5,716.

The real test of strength of the fullblood market in the U.S. however came to surface at the DeMier-Sudbrink sale at Miami, Oklahoma in 1979. Here they sold 73 pure French bulls for an average of $5,885 and the 250 fullblood females brought $4,905. Advertising prior to the sale spoke of "100 heifers by five sires out of 16 females, all born within 60 days, were in one pasture" for the buyers selection. No one guessed that many fullbloods would add up to such a higher dollar event. The widespread demand for them was demonstrated by the fact there were 113 buyers from 19 states and Canada.

The Sudbrink event added new impetus to the fullblood market. Embryo transplant became an "on the farm" procedure and available fullbloods increased rapidly.

Bob and Mary Vantrease were partners-managers of the Ellis-Vantrease Limousin herd in Blanchard, Oklahoma. When it dispersed in 1983 it averaged over $6,800.

Chapter 6
Conclusion

As the Limousin breed moves into 1987, the efforts made to secure the breed a position of importance in the United States during the years 1972-1978 seems even more important than it did at the end of this period of time. It makes the struggles to overcome the element that did not believe in us or our cattle even more meaningful.

Those of us who participated in the breed's introduction and development oftentimes could not understand why it was taking so long. Fat was the filthiest word in the beef industry. It took, and still takes, 2.5 times more energy to produce a pound of fat as it does to produce a pound of quality red meat...yet ten years later our grading system still encourages excess fat production. The consumers have made it clear they do not want the fat and for the first time, it seems the packers and the beef industry are listening. Had not the breeders and planners of the Limousin breed done so well in laying down a sound foundation the breed would not have prospered and expanded as it did from 1978 to the present day. We would not be ready for the great demand and responsibility the beef industry is looking to us to furnish in the last decade of this century.

The wise leadership of the breed has protected the coveted high cutability of our prepotent heavy muscled breed with its easy calving reputation. They have installed the most sophisticated performance analysis to seek the bloodlines with the highest economically important traits. The juniors, the ladies and the men of the breed are involved.

The efforts of the era seem most worthwhile.

Part Three 1978-1986
The Acceptance of the Breed

by Dr. Greg Martin

Introduction

It has been an honor and a privilege for me to be the executive vice president of the North American Limousin Foundation the past nine years. Before giving my recollections of some of the important events that have happened to the Limousin breed in the last nine years, I feel it is fitting to give you a brief background on myself as a form of introduction to this segment of the book.

I was born and raised in the Sandhills of Nebraska and was involved in the family commercial cattle operation throughout high school and during most of my college years. We had a typical commercial program of 150 to 350 straight-bred Hereford cows that were bred back to Hereford bulls with the calves being sold in the fall at the time they were weaned. In the late 60's we began keeping the heifers over, wintering them and selling them in the spring of the year. In an attempt to increase the pounds we had to sell, we started doing A.I. work in the late 60's using Hereford bulls. My brother and I had a small group of Black Angus females, the result of 4-H projects started when we were relatively young. My dad bought us black cows to make sure there was no confusion as to which cows belonged to us boys. In the early 70's we started using Simmental semen on some of the Hereford cows and Limousin semen on some of our Angus cows. Our crossbreeding experiments really opened our eyes as to what heterosis can do for a commercial operation. It also pointed out some of the problems. We did experience some calving difficulties with the Simmental bull, but found the Simmental-sired calves weighed more at weaning time. The Limousin-sired calves came easy and were very vigorous at birth, but tended to

Greg Martin has served as Executive Vice President of the North American Limousin Foundation since 1978. His leadership during this tenure has extended beyond the breed. He is past president of the U.S. Beef Breeds Council and has served six years on the board of the Beef Improvement Federation. During the 1986 International Limousin Council Convention in Limoges, France, Martin was honored by the French Minister of Agriculture as a Chevalier de l'Ordre du Merite Agricole for his contribution to the Limousin breed.

wean-off at lighter weights. My dad ended up using Limousin-cross bulls on our entire commercial herd, due to the calving ease and the calf's livability. At that time my dad also had a job off the farm, so calving ease was a major factor in using Limousin bulls. I had a younger brother involved in the 4-H program while I was in college, and the tremendous carcass merit of the Limousin-cross steers was very impressive. As time went on we saved back some replacement heifers. We found the nutrition in our part of the country was somewhat limited when it came to maintaining the large-framed, heavy-milking Simmental cows so they would rebreed quickly after calving. Limousin cows were smaller in stature and overall size, but seemed to be better adapted to the kind of nutritional environment we could provide.

After completing my B.S. Degree at the University of Nebraska in Animal Science and Agriculture Education, I made the decision to work on a Masters Degree. I chose Purdue University as there would be an opportunity to assist in coaching the livestock judging team. My thesis work at Purdue dealt with the growth and development of red meat animals. My thesis project was with lambs. I slaughtered crossbred lambs at 10 pound intervals from birth (10 pounds) up to 150 pounds. The lambs were frozen in a standing position and then cross-sectioned at several points, like the stifle, 12th rib and 6th rib. Colored photographs and slides were taken of each cross-section at each weight. The slides could then be used to show the muscle and fat development and changes from birth through 150 pounds live weight. This procedure had been done previously with hogs and was an excellent tool in teaching students how meat animals grow and develop.

Coaching the livestock judging teams and working with meat evaluation teams was truly a great experience. While at Purdue I worked with Dr. Roger Hunsley, assisting him in coaching a national champion livestock team. We also spent a great deal of time in the coolers looking at carcasses and cutting up carcasses in the meat lab. It didn't take long to see the Limousin-cross cattle were exceptional in their ability to hang high-yielding, very lean carcasses with exceptional ribeye size.

After the completion of my Masters Degree, I had the opportunity to go to the University of Missouri as a full-time instructor. I was in charge of coaching the live animal judging teams, working with the meat animal evaluation team for the Omaha contest and teaching beginning courses in livestock judging and

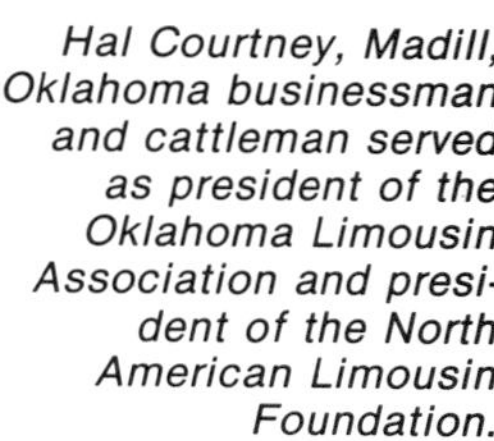

Hal Courtney, Madill, Oklahoma businessman and cattleman served as president of the Oklahoma Limousin Association and president of the North American Limousin Foundation.

Don Faidley served two terms as NALF president. Dorothy, presently president of the Iowa Beef Industry Council, was president of the Limouselles 1982-1984. They farm and raise Limousin at Colfax, Iowa. The couple were made honorary members of NALJA in 1986.

carcass evaluation. After a short while in Missouri I assumed the responsibility for scheduling the livestock center and the animals needed for classroom use throughout the department of animal science. During five years at the University of Missouri I managed to work in enough classes and to gather enough data to complete my PhD in beef cattle management with collateral fields in business administration and agriculture education. The job at the University of Missouri gave me an opportunity to work with students from different backgrounds, and to work with purebred breeders in the state of Missouri. Not only did we work a great deal with beef cattle people, but a little with swine people. I was impressed that swine people were very progressive, were looking forward to the future and trying to change their product. Swine people were able to change their animals so fast genetically because of the short generation interval and the multiple birth aspect. If their goals were wrong they achieved them just as quickly as if they were right.

While at the University of Missouri I witnessed a real change in the swine industry. At one time, the commercial swine people had a very simple, three-breed rotational crossing system that worked exceptionally well. White Yorkshire hogs were used in the beginning because of their excellent mothering ability. These sows were bred to Duroc boars to produce F1's that were tougher, more durable and would grow faster. The F1 York-Duroc crosses tended to have too much fat on them. These animals were bred to Hampshires, which at that time were a heavy-muscled, lean breed, producing a very desirable carcass. The rotation was then started over using the three breeds in the same order. Then problems began. The different breeds of swine decided they were going to make their breed be all things to all people and make all breeds look alike. The Yorkshires tried to make their hogs tougher, more durable and grow faster. In doing this, they lost much of their mothering ability and milking ability. The Duroc's were determined to make their hogs leaner, and in doing so made their hogs grow slower, losing some of their durability. The Hampshire's were determined to make their hogs better mothers, they lost their muscle expression and leanness. The next thing, the commercial man no longer had three breeds that complemented each other in a rotational system. Therefore, the commercial man had to look elsewhere to find specific lines or breeds that could be used in a rotational program to complement one another. Many of the commercial hog producers ended up using hybrid seedstock produced by one of the large corporations that were producing five or six different lines specifically designed for crossbreeding. This is one observation I will never forget and one I think is important to the cattle industry and to Limousin. We definitely need to maintain our strengths and decide where we fit into a rotational crossbreeding program.

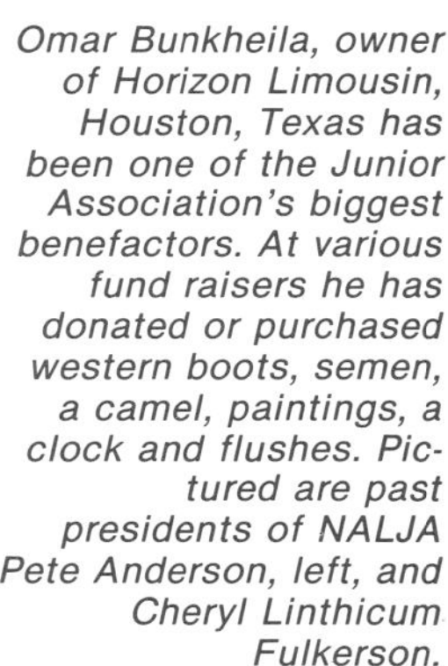

Omar Bunkheila, owner of Horizon Limousin, Houston, Texas has been one of the Junior Association's biggest benefactors. At various fund raisers he has donated or purchased western boots, semen, a camel, paintings, a clock and flushes. Pictured are past presidents of NALJA Pete Anderson, left, and Cheryl Linthicum Fulkerson.

After five years at the University of Missouri, it became apparent when I finished my PhD I would need to look for another job. The university had made a decision to maintain my current position as an instructor and not to elevate it to that of an assistant professor. Throughout my five years at Missouri I spent a great deal of time in the summer judging livestock shows around the state of Missouri. In August 1977 I was at Springfield, Missouri, judging steers and hogs at the Ozark Empire Fair. I remember picking the grand champion steer on a Wednesday afternoon that was a Limousin-cross. The steer belonged to one of the O'Brien boys of Pineville, Missouri and was a big-framed, thick, meaty steer for his time. The next morning after judging a hog show, I went back to the arena and watched the 1977 National Junior Limousin Heifer Show. I had some Limousin cattle in Nebraska at that time and I was interested in the Limousin breed. I read the *Limousin Journal* every month to keep abreast of what was currently going on throughout the country. At the junior show an old friend of mine, Craig Schrader, who at that time was working for the *Limousin Journal*, stopped by to visit. Craig and I had met several years before at Purdue. Craig told me Bob Vantrease had just resigned and the Foundation would be looking for a new executive vice president. I didn't really think I was interested in that type of position and felt I wanted to stay in the university atmosphere. After thinking about it for a month or so, I decided to put in my application and sent it to Dale Runnion, the chairman of the search committee. A Sunday afternoon in October I was watching a football game when the telephone rang. It was Dale Runnion from the *Limousin Journal* calling on behalf of the search committee for a new executive vice president for NALF. Runnion told me they had several people lined up for interviews that coming Tuesday. He said he was reviewing all the resumes one more time and decided I deserved an opportunity to visit with the

John Bruner, Winfred, South Dakota served NALF as chairman of the Technical Committee from 1980 to 1986. Eileen, an active Limouselle, served as the organization's president. The Bruners were recognized as Seedstock Producer of the Year by the South Dakota Beef Cattle Improvement Association.

John and Jackie Spivey and their CMC operation at McDonough, Georgia were one of the Southeast's early herds. The leadership it provided did much to establish the Limousin breed superiority in this section of the United States. The couple were recently honored by the South Eastern Breeders Association for their leadership.

selection committee. He said if I could be there at 2:00 p.m. on Tuesday, they could work me in. I checked with the airlines and made arrangements for someone to teach my classes on Tuesday.

With arrangements made, I started to think about the Limousin breed and what my thoughts would be in regard to questions the selection committee might ask. Having been in the state of Missouri for five years where they had a very strong performance testing program, I knew the Limousin breed must concentrate on a more complete performance program. I also felt the breed had to concentrate on structural correctness, a little smoother muscle pattern and yet still maintain their thickness and carcass superiority. When I arrived in Denver I was introduced to the search committee. Dale Runnion was chairman, Carlton Noyes was the current NALF president, Bill Dameron was the vice president and past presidents Burwell Bates and Floyd McGown were also present. After a two hour interview, I was on my way back to the airport. I thought the interview had gone well and I was impressed with the gentlemen I met that day. About 10:30 p.m. that evening the telephone rang; it was Carlton Noyes. Carlton indicated I was the search committee's first choice for the position and if I was interested I needed to be in Denver the following Sunday to meet with the entire board. After a couple days of soul searching and long discussions with my wife, Pat, the decision was made to go to Denver and meet the entire board.

My wife and I made arrangements to get to Denver on Sunday morning. After lunch with the board, we spent a long afternoon answering questions and discussing the Limousin breed with the entire board of directors. After all bases had been covered, the board announced the job was mine. The next day we spent in the office looking over the operation and meeting the staff. The conditions had been set by the board that my first day on the job would be at the annual meeting in Louisville, Kentucky, November 16th and that I would be in Denver on a full-time basis starting January 1, 1978. I would never have guessed six months earlier that I would be working for the North American Limousin Foundation. Fate has a strange way of taking us in new directions. I am thankful the search committee and the board of directors had faith in me and gave me the opportunity to be involved with the North American Limousin Foundation.

I have chosen to divide my section of this book into six chapters, rather than cover the past nine years chronologically. These chapters reflect some important areas in the history of the breed since 1978, and I hope they will be informative and interesting. These are simply my recollections and opinions as to what I think the important events have been for us to remember in the years to come.

Darrell Wiggins, Houston, Texas purchased the Deux Amis herd which included Nordic Debutante and Nordic Debonnaire. The man and the cattle created so much excitement at the turn of the 70's that some refer to it as the "Wiggins" era.

Chapter 1
Acceptance of Limousin by the Beef Industry

In the fall of 1977 the cattle business was still relatively depressed. We had just been through the big crash of 1974-75 and people were very cautious about building back numbers and getting involved again with cattle. The people involved in agriculture and livestock were leaning toward the conservative and were reluctant to use some of the new breeds. To say that Limousin was widely accepted as a viable breed in the beef industry would have been stretching the truth. There were small groups of commercial breeders from the very strong Limousin states that had used the bulls and were satisfied with them, but Limousin bulls were not widely used across the United States in commercial herds.

In the fall of 1977 after the announcement that I had taken a job with the North American Limousin Foundation, I had an opportunity to visit with a good friend who was also a judging team coach. This gentleman wished me the best of luck on my new job, but told me he didn't really believe the Limousin breed had anything to contribute to commercial beef production, and he didn't believe the breed would ever be widely accepted. As I started traveling for the Limousin Foundation in early 1978, I found that when the word Limousin was mentioned most people said, ''They're cars, right?'' We definitely had a large scale educational problem on our hands to convince the cattle industry of the merits of Limousin, and of the fact Limousin could contribute to more profitable beef cattle production in this country.

A couple of years earlier the board established a research project with Oklahoma State University (OSU) and Dr. Dick Frahm to gather all the information they could on the Limousin breed. In the spring of 1978 an OSU bulletin was published which was the result of this two-year investigative process conducted by Dr. Frahm and his associates. The bulletin was a first of its kind in the beef cattle industry. Not only did it expound the true merits of Limousin cattle, but it pointed out the areas in which the breed had some weaknesses and needed to improve. The bulletin was made public and was mailed to anyone who requested research information on Limousin. Many of the old-time breed associations couldn't believe we would publish an official research bulletin that wasn't 100 percent pro-Limousin. The board's philosophy was that we needed to know the strengths and weakness of our breed and we needed to sell our breed to the commercial man based on what it honestly could do to help them.

We did take the most positive aspects of the OSU report—the calving ease of Limousin bulls and the profit per cow bred to Limousin bulls—and used it extensively in our first major advertising campaign. The campaign was built around the headline, ''The Word is Limousin.'' The general idea was to get people familiar with the word Limousin and make them connect it with the cattle business rather than cars. The ads had some attractive art work of bulls and baby calves, which allowed them to run very effectively in black and white as newspaper ads. The ads were developed with the help of Dale Runnion, who was then publisher of the *International Limousin Journal.* These ads were used for approximately two years as the Foundation's ad budget grew from nearly nothing to over $75,000.

THE WORD is Limousin

We had our first display booth at a National Cattlemen's Association (NCA) Convention in 1979. At that point, most of the people who stopped by the booth were asking questions. What is a Limousin? What do they do? Why should I use them? Where can I see one? It was obvious that most of the people attending the NCA Convention did not know Limousin was a breed of beef cattle.

Limousin bulls were selling well to the commer-

The adaptability of Limousin has been given a real test in North America. This herd is pictured just outside the Calgary city limits, virtually in the foothills of the snow capped Rocky Mountain range in Alberta.

The great numbers of Limousin in the state of Florida attest to the fact that palm trees and Limousin do coexist.

Denver's crisp winter air makes for frisky bulls when they are turned loose in the sand arena at the National Western Carload bull show. Andy Rest snapped this fellow with all four in the air on his romp through the ring caring naught whether the judge gets an appraising look at him.

Beauty is where you find it, like this rare snow blanketing west central Texas and illuminating the consistent quality of North American Limousin.

Juniors lined up for a class are pictured at the 1984 National Junior Heifer Show in Rapid City, South Dakota. Junior shows are held in conjunction with most major shows in the United States. In addition, most states hold state and regional shows during the year giving juniors several chances each year to participate.

Bloodlines containing three-eighths Brahman blood have been popular for a long time in the Gulf Coast states and the Southwest region of the United States. Brahmousin with their five-eighths Limousin blood are adding thickness to the hardy breed and extending the area of acceptability of the "eared" cattle.

One of the greatest benefits of the NALF junior program is to see the smile of a junior who has been victorious. Vernon McKown (pictured left) and his brother Richard started into a junior program, their father got interested and today McKown Limousin, Norman, Oklahoma is one of the country's most successful breeding herds.

Nordic Debutante became well known after this great picture (below) by Vivian Sirratt taken at Nordic Farms, Campbellville, Ontario, Canada was published as a cover on the Limousin Journal. After going to Deux Amis, Luling, Texas, Debutante sold to Wiggins, then Bingham and later to Bishop Boys. The prolific producer was heavily transplanted all her life. In 1979 three fullblood sons (transplants by Carnaval) won the pen of three bull championship in the Denver yards.

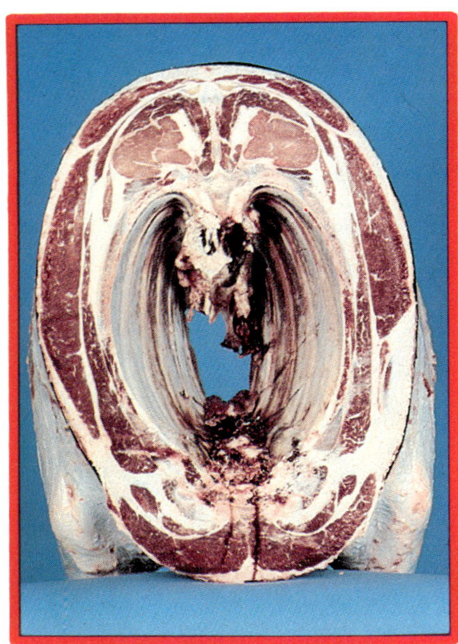

Lots of saleable, bright lean beef with little fat trimming in this section.

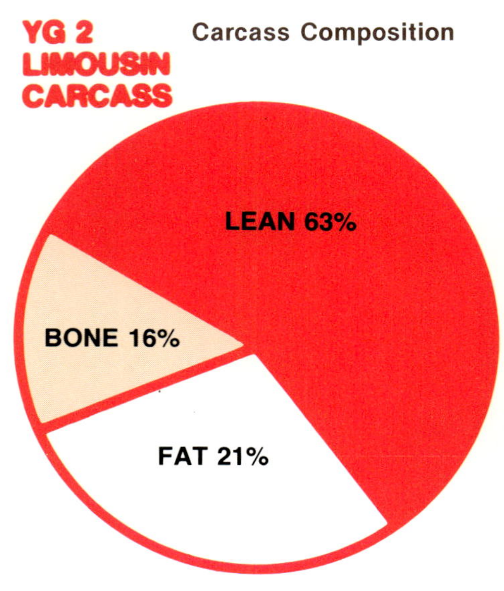

The Limousin carcass produced 20 percent more lean than the overfat, light-muscled carcass.

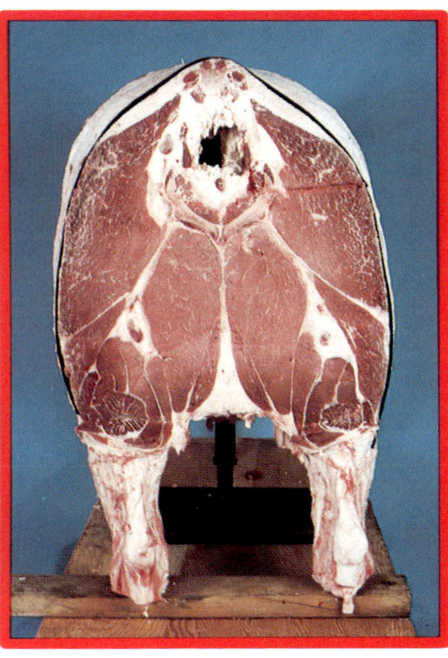

The big, thick muscles make for a lot of saleable product.

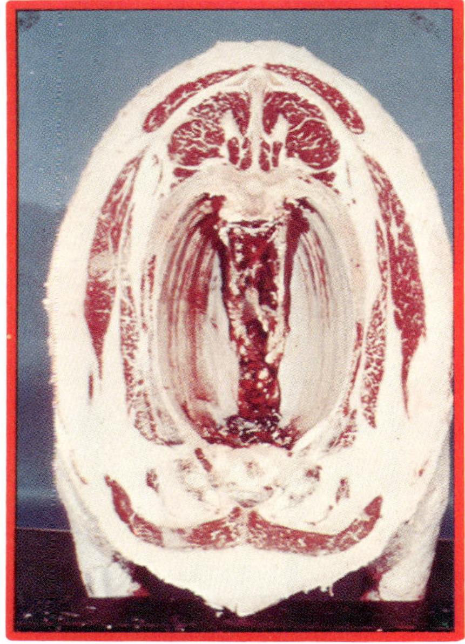

Note the seam fat and the excess fat in the brisket.

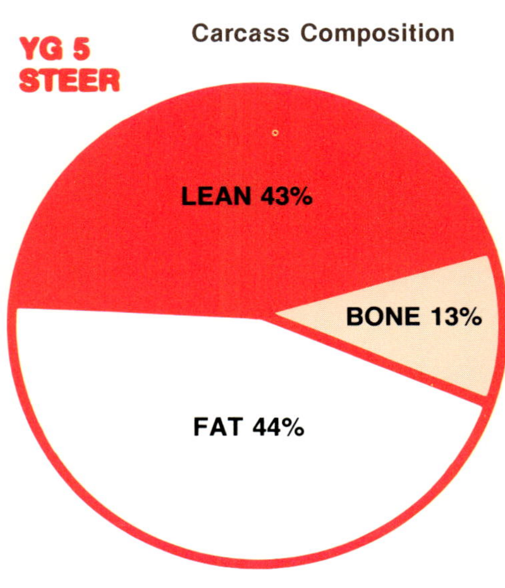

The chart shows an unbelievable 44 percent fat in the five yield grade carcass.

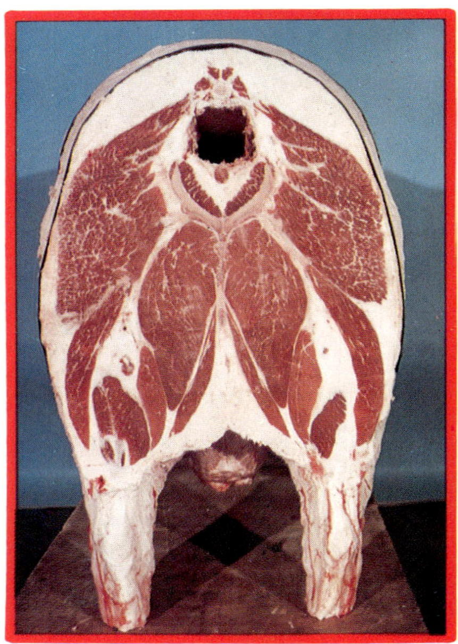

This cut shows the extreme excess fat over the rump and in the twist.

LIMOUSIN BEEF IS GENETICALLY TRIMMED.

Pictures and graphics depicting the cross cut sections of two steer carcasses frozen in a standing position have been some of our breed's most useful materials for promotion. The choice Limousin carcass is compared to an overfat carcass both in the crosscut picture and accompanying graphics.

Grand champion carload at the 1986 National Western Stock Show was exhibited by Yackley-CMC, Onida, South Dakota. The Yackley firm exhibited Denver carload winners in 1982-1983-1984-1985 and 1986.

Grand Champion pen of three bulls honors at the 1987 National Western Stock Show went to Symens Brothers, Amherst, South Dakota. Symens won this division of the Denver contest in 1980, 1982, 1983, 1984 and 1987.

Spitz Navajo (X 747 out of CMC Miss Snowflake) was the first recipient of the Limousin Triple Crown award. The Triple Crown was authored by American Cattle Services who donate the trophy given to that bull or female with consecutive grand championships at the All American Futurity, the American Royal and the National Western.

Spitz Special Effort (X Sonic Jet out of CMC Miss Crystal) was also from the Norman, Oklahoma Spitz string and became the second Triple Crown winner with her win at the 1986 National Western. The two winners were out of full sister flushes.

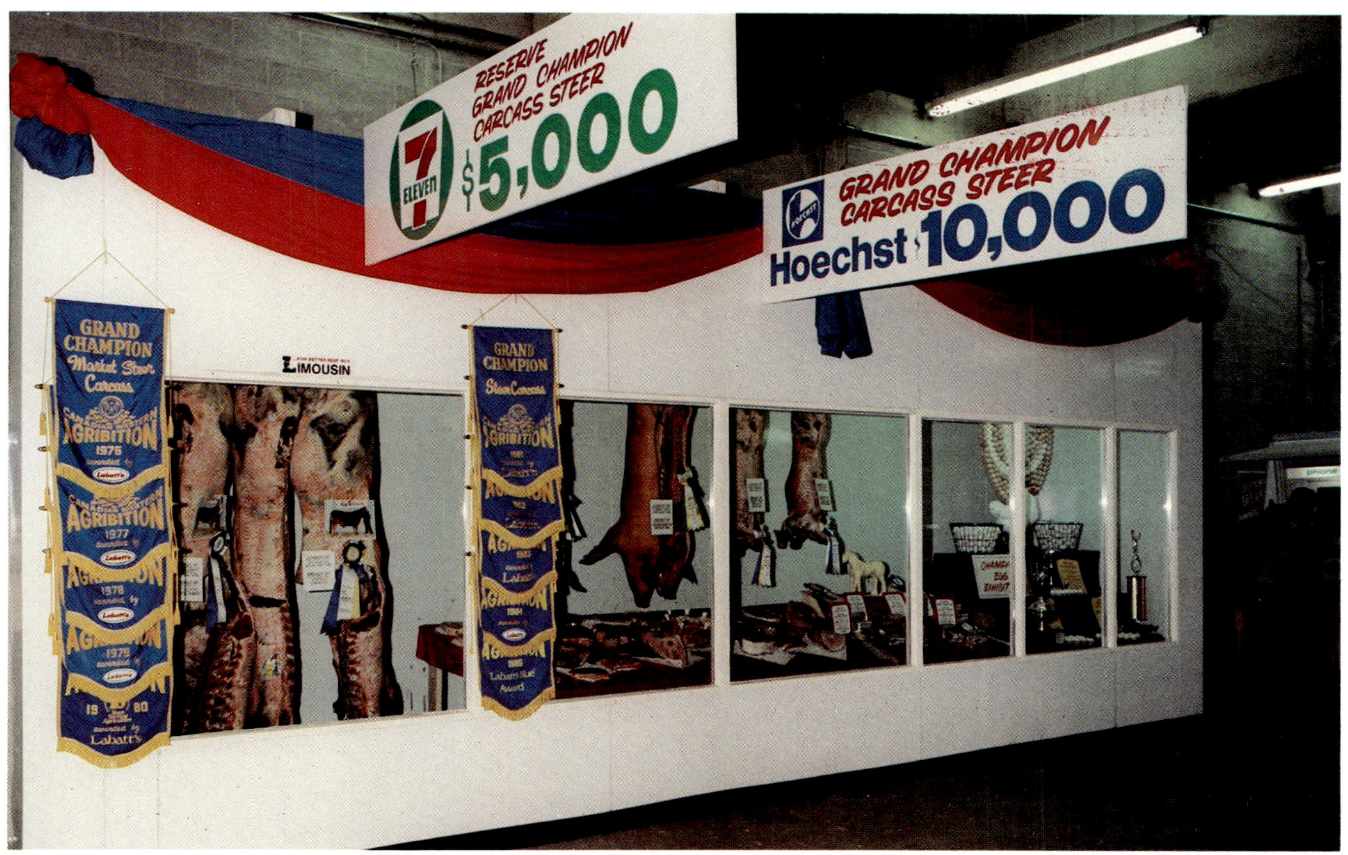

Limousin carcasses have won the grand championship at the Canadian Western Agribition in Regina, Saskatchewan for eleven consecutive years. In addition to the grand award, the reserve and next fifteen carcasses wore a Limousin hide, in 1986, before they were put on the rail for the carcass show.

Some ranchers still put the brand on their calves "the old fashioned way—they work for it". These Davis mountain raised calves give their Texas cowboy owners a good "rassle" in the process.

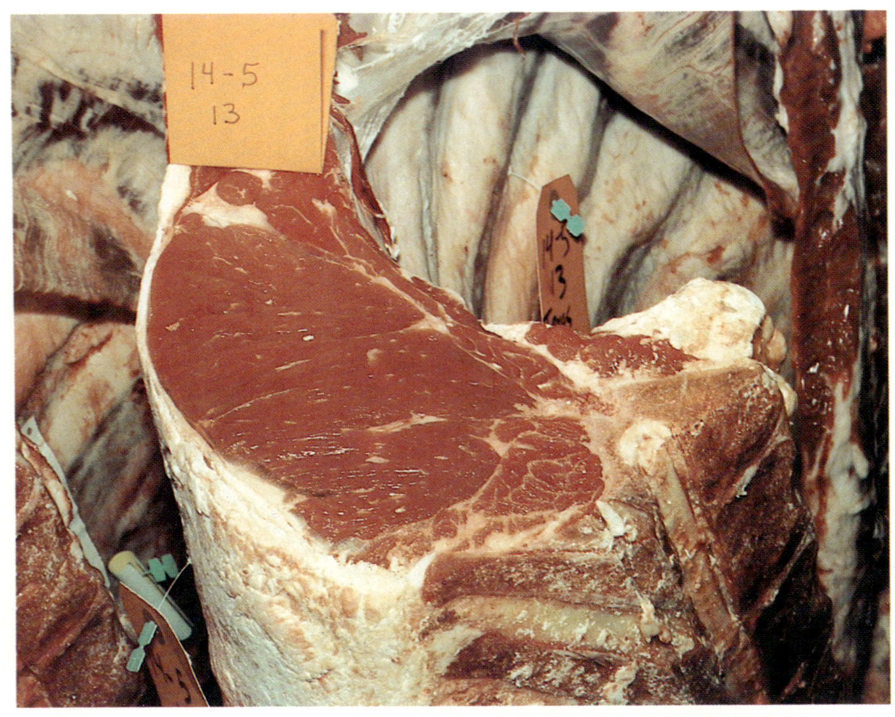

Since World War II, Americans' taste for highly finished beef has exceeded that of most any other country. The changing demand today made this carcass with its large ribeye, slightly marbled with a thin fat cover, a high priority in the trade. A Limousin bull genetically trimmed this carcass. Pictured below is a part of the old Denver Stock Yards now used during Stock Show week for carload and pen, purebred and commercial cattle entries. At the 1987 National Western 18,983 head of cattle sold at auction for a total of $10,806,755. The totals include sales held in the yards and purebred auctions "on the hill".

Historic Pantops overlooking Charlottesville, Virginia is the home of Worrell Land and Cattle Company. Their annual sale since 1983 has been a breed pacesetter, each one featuring the get of Seven Forty Seven. The beautiful farm owned by Gene and Anne Worrell is decked out for sale day in the above picture.

This pen of Limousin x Angus steer calves bred by Jerry Adamson, Cody, Nebraska won the grand champion pen of feeder calves trophy at the 1974 National Western. The championship is coveted by all breeds because most times they are purchased by experienced feeders who have them exhibited later in interbreed competition.

FOR SALE BY
THE DENVER
LIVESTOCK MARKET
PEN of 5
FEEDER CALVES
Angus & Limousin Cross
Consigned by
JERRY & DELORES ADAMSON
Cody, Ne
HANGE
YARD

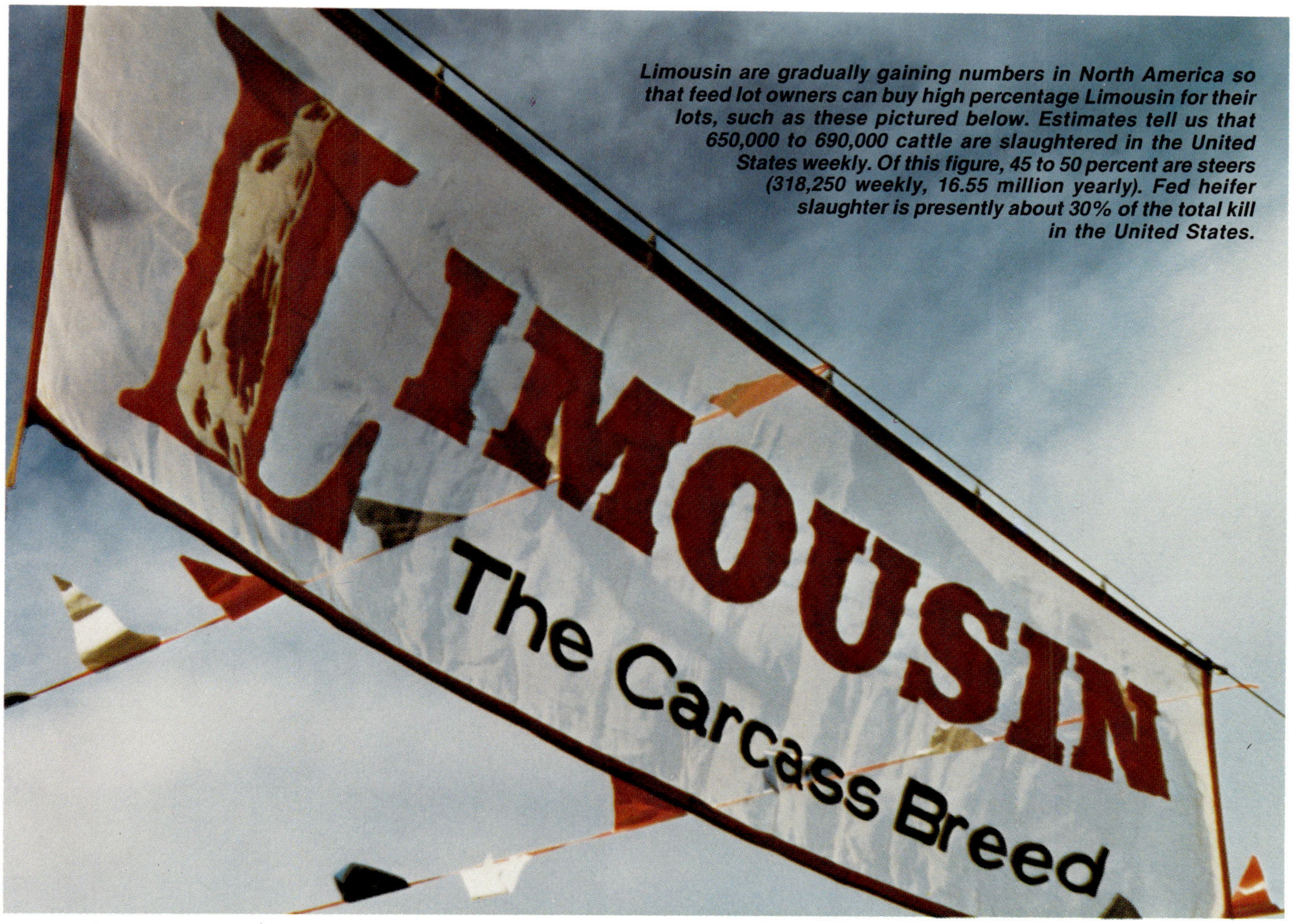

Limousin are gradually gaining numbers in North America so that feed lot owners can buy high percentage Limousin for their lots, such as these pictured below. Estimates tell us that 650,000 to 690,000 cattle are slaughtered in the United States weekly. Of this figure, 45 to 50 percent are steers (318,250 weekly, 16.55 million yearly). Fed heifer slaughter is presently about 30% of the total kill in the United States.

Sirratt

cial cattlemen in the areas where we had a large number of Limousin breeders who had worked hard at promoting and selling their product. States like South Dakota and Oklahoma had large numbers of Limousin breeders, and bull sales to commercial cattlemen were good.

My past relationship with the university people turned out to be a positive aspect in my new position with the Limousin Foundation. The *Limousin Journal* and NALF had been sending many of the university people complimentary subscriptions to the *Limousin Journal* for several years. This program was expanded to make sure all university personnel who worked with beef cattle in any way, whether teaching, research or extension, were put on a list so they could receive the magazine. Many of the university people were pleased with the approach we had taken with our OSU bulletin. The positive aspects were used from a promotional standpoint, and those traits needing improvement were identified for breeders to work on. Academic people definitely identified with the bulletin. I also stayed as involved as I could with my old colleagues by assisting with judging contests, speaking at various university functions and attending industry-related meetings where large numbers of university faculty were involved. Other research programs were started, one with the University of Georgia and another one with OSU, to gather additional information on Limousin. A previous research project had been completed at Texas A & M and our *Sire Summary* was starting to gain recognition throughout the industry. This was definitely a time when most universities were promoting performance testing programs so our *Sire Summary* and our basic performance program were generally well received.

After a couple of years using, "The Word is Limousin" in our advertising program, we switched gears to a new program called, "Discover the Limousin Extra." The ads in the program were again developed with the assistance of Dale Runnion, and the response to the new ads was excellent. The ads featured large pictures and were primarily in black and white with one added color. The advertising budget was increased and ads were directed towards selling bulls to commercial cattlemen and increasing the value of Limousin feeder calves. The number of Limousin sales across the country began to grow, with much strength developing in the Southeast and

in Texas. At the NCA Convention in January 1982, it appeared attitudes had changed drastically. We were now in a situation where people knew what a Limousin was and a good many of those who stopped by our display had either used some Limousin bulls or fed some Limousin calves. Overnight we came of age. People thought of us as one of the stable breeds in the new beef complex. We were no longer the new kid on the block or the new exotic with the funny name that sounded like a car. We were Limousin *beef cattle*. We were known for calving ease, good growth rate and cattle with thickness and muscling that feeders and packers loved. This acceptance was a result of the previous 10 years of work put in by the Foundation and the breeders to develop an honest and sincere image with the beef cattle industry as to where our cattle fit in and what they could do to help the commercial cattlemen. There is no doubt state associations, with their efforts in putting on shows, exhibits and sales, had a great deal to do with the widespread acceptance of Limousin. Since 1982 the breed has continued to expand its area of influence to all corners of the United States. More and more bulls and females are showing up in other parts of the country, with areas such as the eastern U.S. and the Northwest showing more activity in recent years. A new advertising campaign was started in the fall of 1983 with more emphasis on the Limousin female and her productivity. More ads were run in four-color and the advertising budget was increased to over $100,000. Bull sales continued to grow in numbers as Limousin earned a solid reputation for calving ease and for feeder calves that would bring a premium. The academic community responded with compliments toward our breed and positive declarations as to what the breed could do to help commercial cattlemen.

The gentleman who told me in the fall of 1977 that he didn't think the cattle would ever work, happened to be judging one of our major state shows in the fall of 1982. After the show was over he told me he realized he was wrong. He was convinced

U.S. Limousin Registrations, Transfers and Memberships
By Year

Fiscal Year	Registrations	Transfers	New Actives	New Juniors
77/78	23,741	18,734	309	158
78/79	25,539	19,492	359	164
79/80	29,989	18,292	472	147
80/81	39,195	27,510	652	259
81/82	42,590	30,725	828	186
82/83	47,814	35,194	875	240
83/84	47,135	34,428	821	401
84/85	41,739	30,624	761	365
85/86	37,746	23,006	423	289

Table showing U.S. Limousin Registrations, Transfers and Memberships by Year is supplied by the North American Limousin Foundation.

Limousin cattle did have a great deal to offer the commercial beef cattle industry.

Registrations continued to increase steadily from 1977 until 1984, and the Limousin breed was now firmly entrenched as the fifth largest beef cattle breed association in the United States and the largest Limousin registry anywhere in the world. In 1984 and 1985 the Foundation experienced slight decreases in annual registrations due to the difficult times in the beef cattle industry and the general problems with our national economy. Even with decreases in registrations, the Foundation was adding 600-700 new lifetime members each year, and more than 5,000 commercial people were buying their first Limousin bull each year.

As the Limousin breed became accepted as part of the beef cattle industry, we as an association became more involved in our related organizations. We became very active in NCA and had a seat on the board of directors with representation on many of NCA's standing committees. We are deeply involved in the Beef Cattle Improvement Federation. I had the opportunity to serve on the board of directors for six years. We work closely with the U.S. Beef Breeds Council; I had the opportunity to serve as their president for 1982 and 1983.

There is no doubt the Limousin breed is accepted as an integral part of the beef cattle industry. However there is still a great deal of work to be done. There are many people who have not been exposed to the breed and do not understand the merits and values of using Limousin bulls in their programs. The challenges ahead are great. As you will see at the conclusion of my part of this book, the board has developed a long-term plan to continue to push Limousin throughout the industry and to gain a larger portion of the commercial bull market.

Leonard and Jerry Wulf and their integrated Wulf Farms operation, Morris, Minnesota consigned the grand champion pen at the 1986 National Western Fed Beef Carcass Contest. The firm has also been consistent winners or contenders in the National Western pen and carload bull show.

Chapter 2
Performance Programs

Prior to 1977, the Limousin Foundation had a good, basic performance program. The first sire summary produced in 1974 was one of the first field data sire summaries published by any breed association. Sire summaries were also printed in 1975 and 1976 and the breeders were being provided with adjusted weights and ratios within their own herd. A sire summary was not produced in 1977 due to some interaction problems that Dr. Benyshek was having between the halfblood data and the three-quarter and seven-eighths data. A new sire summary in 1978 dealt with only three-quarter and greater cattle as a way of solving the previous problems. In missing the one year in 1977, it became obvious to the board of directors that there were a large number of Limousin breeders who were faithfully studying and using the sire summary each year.

The sire summary continued to grow and expand in 1979, 1980, 1981 and 1982. In 1982 a new and improved model was used, which took into consideration the sire birth year effects. This new model was developed assuming that the group of bulls born in 1977 may be genetically different from the bulls that were imported and were used in 1968. This model was designed to compensate for the genetic trend or the genetic drift as improvement was made in the traits each year and new and better sires were added. This was necessary in order to make accurate comparisons between young bulls and old bulls. In 1983, the procedure was again improved to use the relationships among the sires to enhance the evaluations. This new method of analysis also increased the number of bulls in the sire summary from 399 in 1982 to 1,109 in 1983.

In 1984, the genetic merit of the dams was used in a national sire summary for the first time anywhere in the world. This new model used the dam's breeding values and adjusted each bull based on the dams to which he was bred. The basic model was still a BLUP (Best Linear Unbiased Prediction), but by using the genetic relationship matrices for both the sires and the dams we produced the first sire summary to have this extra added degree of accuracy. No longer could a bull's breeding value in the sire summary be distorted because he was bred to better cows than another bull. The new model also increased our ability to evaluate more sires and the number rose to 2,728 bulls listed in the *1984 Sire Summary*. This sire summary was definitely the model of the industry and the first one to achieve this degree of accuracy and completeness. Dr. Larry Benyshek deserves a great deal of credit for all the time and effort he put into developing and implementing this new and revolutionary sire summary.

Symens Brothers, Amherst, South Dakota are numbered among the great breeders of Limousin in the United States. Herman, pictured left, served on the National board and was NALF president. His brother John is president of the South Dakota Limousin Association. Brothers Paul and Irwin are also members of the firm.

The *1985 Sire Summary* was again a year of changing and improvement. NALF produced the largest sire summary ever printed in history when 3,742 bulls were published in the *1985 Sire Summary*. The *1985 Sire Summary* considered all available information in calculating its expected progeny differences (EPD's). In the past, the ratios from the contemporary group analysis of progeny only were used to come up with the sire summary EPD's. In 1985, the analysis used all of the individual's progeny, pedigree information on all of his relatives, and his own individual performance within his contemporary group. A major change in the *1985 Sire Summary* was from maternal ability to true milking ability. Previous to this time, maternal ability, as measured in sire summaries, was a combination of growth rate and milk production that the bull's daughters contributed to the weaning weight of their calves. This maternal value was very highly correlated to growth rate to weaning time and was not a measure of true milking ability. In 1985, the true milking abilities were calculated which allowed breeders to select bulls for pure milking production rather than a combina-tion of milking ability and growth rate. This turned out to be a very important tool in that it was possible in the past to select bulls that had high maternal values that were actually exceptionally high in early growth rate and negative in milking ability. It's important the traits of early growth rate and milking ability be kept in balance to produce calves with the largest amount of weaning weight and still keep the cows productive under their current management system. The switch to printing the true milking ability was a first for the Limousin breed, as this was the first sire summary ever published that listed pure milking ability.

In 1986, the sire summary calculations remained much the same as in 1985. As in 1985, it was produced with the new "Reduced Animal Model," which adjusts sires for the superiority or the inferiority of the dams they were mated to, reduces the effects of non-random mating, accounts for the genetic trend of young bulls and uses the most complete, accurate and advanced sire evaluation techniques available. The format of the *1986 Sire Summary* was changed slightly to reflect current changes within the breed. In order for bulls to be listed in the *1986 Sire Summary* they must have had at least five progeny recorded in the last three birth years. Trait leader accuracies were raised to .70 so bulls that were on the trait leader list would have a higher confidence level. In the general listing bulls with accuracies of .5 or greater were still printed. An exciting new addition was the "Young Sire Prospect" list. In order for a young bull to qualify for the prospect list he must have been born in or after 1982, he must have had EPD's in all four traits measured and accuracies must be below .7 as those bulls with .7 or above would be in the regular trait listing. These young bulls had to have valid contemporary group data at weaning and yearling and must have had an individual birth weight. The milking ability EPD in this section is strictly a pedigree evaluation on these young bulls. The purpose of the young sire prospect list was to identify young bulls which are potentially strong in the traits evaluated. These young bulls are likely candidates for a sampling program where a breeder would use them sparingly to help gather additional information to determine their true breeding values.

With the help and guidance of Dr. Larry Benyshek and his staff, the Limousin breed has been

Wayne Vanderwert (left), NALF Director of Research and Education, has been a Limousin enthusiast since 4-H days in southwestern Minnesota. Auctioneer Carroll T. Cannon, Tifton, Georgia served as field representative in the Southeast for the Limousin Journal *before becoming a full time livestock auctioneer and sale manager.*

at the forefront of sire evaluation since 1978. Dr. Benyshek's persistence in developing new, better and more advanced models in which to do sire evaluation has been a real asset to the Limousin breed. The increased size, quality and accuracy of the *Limousin Sire Summary* brought a great deal of attention to the Limousin breed's performance program and definitely helped with the acceptance of our breed in the academic and extension fields. The NALF Board of Directors needs to be commended for having the willingness and the foresight to support Dr. Benyshek's new programs in developing the most advanced and complete sire summary in the industry today.

Along with all the work being done on sire evaluation, Dr. Benyshek was also developing individual EPD's for the Limousin breed. In conjunction with the *1985 Sire Summary,* the breed developed its first-ever estimated progeny differences (EPD's) for all animals in the herd book that had valid data. In the spring of 1985, Limousin breeders were all mailed inventory sheets listing every animal on file they currently owned and the EPD's available for those individuals. This was the first time in the history of any beef cattle breed association the members were all provided with individual EPD's on every cow in their program. There are some cattle that did not qualify for EPD's if they did not have individual performance records within a contemporary group or if data was missing on their sire or dam. The new EPD's were a way of ranking all animals across the breed rather than just within the herd where they were raised. The ratios and adjusted weights had

been good indications of how animals ranked within their individual herd, but did not rank them across the breed. The new individual EPD's were exactly like the sire summary EPD's as they were a comparison across all herds and based on a breed average of zero for each trait. The new EPD's gave breeders another excellent tool to use in culling and selection and in making planned matings. A good example would be someone who has been having difficulty with birth weights that are too large in his Limousin herd. Too many times in a situation like this the sire is blamed for all the heavy birth weights. In most cases, if the breeders go back and study the EPD's on their cow herd, they will find out the cows that are having the extremely large calves also have high EPD's for birth weight. Therefore, the problem of the calf that weighs over 100 pounds is a combination of a cow with a heavy birth weight EPD and a sire with a heavy birth weight EPD. People can look at the EPD's and if they wish to use a sire that will improve the weaning and yearling weight of the calf crop, but has a heavy EPD for birth weight, then that bull could be mated to cows that have negative EPD's for birth weight in order to balance out this particular trait and keep birth weights from getting too large.

The important thing in using and looking at EPD's is for breeders to realize it is not likely that an individual animal is going to excel in every trait. It is extremely difficult to find a bull that sires very light calves at birth and yet has exceptional growth rate at weaning and yearling time. There is also some indication that milking ability is somewhat negatively correlated with growth rate. Those animals that have an extremely high EPD for growth rate are probably

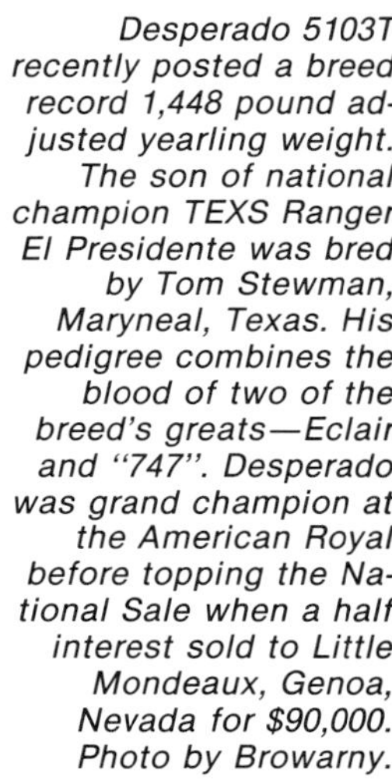

Desperado 5103T recently posted a breed record 1,448 pound adjusted yearling weight. The son of national champion TEXS Ranger El Presidente was bred by Tom Stewman, Maryneal, Texas. His pedigree combines the blood of two of the breed's greats—Eclair and "747". Desperado was grand champion at the American Royal before topping the National Sale when a half interest sold to Little Mondeaux, Genoa, Nevada for $90,000. Photo by Browarny.

not going to be at the top of the list in milking ability. EPD's should be used in mating individuals to come up with the best of both worlds or to produce an animal that is a compromise between the sire and the dam. Cows who lack in performance should be mated to high growth rate bulls. Cows who are low in milk production should be mated to high milk production bulls if you wish to keep replacement heifers that will have more milk than their mothers. The availability of expected progeny differences (EPD's) on both sires and individual cows is the single most important tool we have to provide predictable, consistent genetics to the commercial cattle industry today.

The new EPD's are only an addition to the existing performance records. Breeders are still provided with weaning and yearling summaries, which ratio all the calves and show the averages by sire. The new dam summaries available to NALF members provide a summary of all the calves out of that particular dam, their individual performance and their averages by sex, as well as a calculated performance, and their averages by sex, as well as a calculated MPPA (Most Probable Producing Ability) on the dam. These "within herd" records are still an excellent way to make decisions within individual programs for keeping replacements or culling cattle who do not quite measure up. However, the EPD's are superior to the "within herd" records and should be a better indicator of future performance

than the individual "within herd" ratios. One must realize the EPD's are not currently available on very young cattle. The EPD's are currently calculated for the entire herd book as part of the sire summary evaluations and are, therefore, only done once a year. NALF's computer files are updated each fall with the newest and most complete EPD's and the accuracies for each EPD. Cattle with individual performance records that have pedigree evaluation information available should receive EPD's in the fall after they have turned a year of age. In the future, we hope to be able to calculate EPD's at an earlier date and provide them on younger cattle as weaning and yearling weights are recorded. It's important to point out the need for a better understanding of the accuracy value that goes with the EPD's. Currently, accuracies can range from about .2 up to .99. Young animals with limited data and accuracies of .2 have a greater opportunity for their EPD's to change either up or down as more information is gathered on their progeny. Bulls that have accuracies of .9 or better are not going to see much change in their EPD's even as additional progeny records are added to their file. Therefore, if a breeder wants to use a "sure bet," he should pick bulls with higher accuracies. If he is willing to gamble on a young bull with lower accuracies, he should do so realizing the bull may turn out to be better or worse than what his EPD's may indicate.

EPD's are going to become more widely used. The only real reason to be in the registered seed stock business is to provide superior *predictable* genetics for the commercial cow/calf operator. Superior is hard to define as different commercial men will need to select bulls based on where they need to improve their herds. The key is predictability in that we need to make sure the bull will do what that particular commercial man needs in his herd. If a bull is selected to increase the weaning weight, he had bet-ter do that or there is going to be an unsatisfied customer. The EPD's are definitely our best bet in predicting the breeding value of our Limousin cattle and therefore, they will become even more important in merchandising cattle in the future.

Fun loving E.J. Bishop, Winters, Texas becomes all business when it comes to promoting Limousin. He served two terms as president of the Texas Limousin Association and saw the state become the largest Limousin state in the U.S. under his leadership. David Hatcher (X Bar H), Cameron, Texas is pictured at right.

Chapter 3
Breed Direction

In the late 1970's, most of the new breeds and many of the old breeds lacked real direction. The new exotics had put pressure on the existing English breeds to increase the frame size of their cattle, as well as the performance. Crossbreeding was practiced by a great many commercial people around the country, but primarily as a means to get the old-fashioned, small framed cattle of the 50's bigger, growthier and more efficient. The show ring influence was very prominent in that it had lead the industry towards the bigger, growthier, leaner kind of cattle and the show ring was obviously more visible and more popular than performance programs. Most of the show ring judges were very critical of cattle that were too thick or too heavy muscled. The general trend for all breeds was to have a flatter, longer muscled animal and to make the females look more feminine. Much of this philosophy came from the problems the academic community had experienced with Charolais cattle. The Charolais cattle that were heavy muscled had a lot of calving problems, a great deal of fertility problems and had low milk production. There was no doubt the Charolais cattle did have the double muscled gene in their population and the extremely heavy muscled cattle had all the

problems associated with double muscling.

The academic people had also been associated with the swine industry. In the late 60's, the swine people were real excited about raising carcass hogs with a very high percent ham and loin. These hogs were very narrow through the chest floor, shallow in the body and extremely heavy muscled down the loin and through the ham. It turned out this same type was also very nervous and stress susceptible. Therefore, the extremely heavy muscled hogs were infertile and were not money makers for the commercial man. Their only real purpose was to win carcass shows.

Prior to my going to work for the Limousin Foundation in 1977, I had been involved in judging shows across the country, both swine and cattle. I was very much in tune to the current trends as were other judges. We were all in the situation of selecting both hogs and cattle that were longer, smoother, flatter muscled and females that were more feminine. In many cases we actually selected animals we thought were feminine that were really over-refined and frail rather than feminine. The major issue was muscling. Everyone was desperately trying to make all breeds look alike in muscle pattern and muscle shape. For

The record of Seven Forty Seven (X Espoir de Carnaval out of Fadette) as a breeding bull is unequalled in Limousin history in North America. The record of his get and grand get at the 1986 All American Futurity attest as to his prowess. At this event, Seven Forty Seven was either the sire or grandsire of 15 of the 18 division champions and reserves. He is owned by Worrell Land and Cattle Company, Charlottesville, Virginia. Photo by Stivers.

some reason, we had all decided feminine females should look like Holsteins and therefore, the flatter and longer the muscle pattern, the better. We talked about how these long, smooth, flat muscled females were going to be more productive, give more milk and be better mother cows than cattle with the extra thickness.

After coming to work for the Limousin Foundation, I had the opportunity to visit with Limousin breeders all over the country and to travel to numerous herds throughout the United States. In visiting with Limousin breeders, it was quite obvious most of them felt the differences in muscle thickness did not have any effect on the fertility or the milking ability of their cows. One breeder I visited with indicated that every year he sorted off the open or barren cows in the fall of the year and the only thing those cows had in common was the fact they were either open or barren. He was convinced the heavy muscled cows in his herd were producing just as well, giving just as much milk and doing just as good a

job as the lighter muscled cattle. He actually felt the thicker cattle were easier fleshing and easier keeping. I began to think about what we had seen happen in the swine industry.

In making all breeds look alike, the swine breeds had lost their identity and their real purpose of supplying genetics for crossbreeding programs. After discussing breed direction with several breeders around the country, it became a topic of the North American Limousin Foundation board of directors. What were our real purposes and goals, and what direction should the Limousin breed take? Soon after the board's first discussion regarding breed type and the direction of the breed, a decision was made to have a national symposium dealing with this subject. The symposium was set for the summer of 1979 and after discussion with different universities, Oklahoma State was picked as a sight and a co-host. The symposium would be called "Crossroads '79" with the real emphasis on getting Limousin breeders and academic people together to discuss the direction of the breed in the years to come.

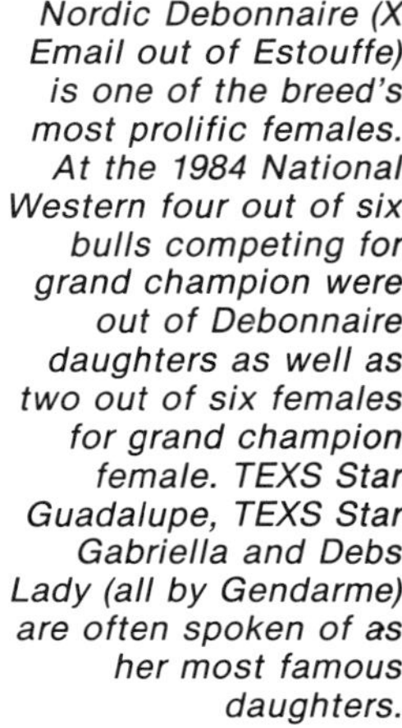

Nordic Debonnaire (X Email out of Estouffe) is one of the breed's most prolific females. At the 1984 National Western four out of six bulls competing for grand champion were out of Debonnaire daughters as well as two out of six females for grand champion female. TEXS Star Guadalupe, TEXS Star Gabriella and Debs Lady (all by Gendarme) are often spoken of as her most famous daughters.

The format for "Crossroads '79" was a full day of classroom talks and discussions by various animal scientists across the country. Dr. Roger Hunsley from Purdue talked on "Limousin muscling, too much or too little." Dr. Hunsley pointed out the Limousin breed is somewhat unique compared to most other breeds of beef cattle. He said that visually the breed appears to be much more muscular than most breeds and yet still possesses a very high degree of reproductive efficiency and functionality. He indicated it would be in the best interest of the Limousin breed to maintain above average muscling as long as it did not limit or hinder reproductive performance. He also pointed out the cattle were longer and smoother in their muscle pattern than they were six years earlier, but that at this point we should not make our Limousin cattle any more angular or any more flat muscled and we should not attempt to make them look like the other breeds in terms of muscle pattern.

Dr. Harold Harper of Penn State University talked on "Frame Size, How Tall is Tall Enough." He pointed out there was not one ideal size or type that would fit production situations. He said it was important to select cattle that were adaptable to specific climates, environments and nutritional availabilities and not to try and make one frame size or type fit all programs. He concluded that frame size had increased greatly and the increase in average frame size appeared to have a positive effect on the beef cattle industry, but we should proceed with caution in increasing the frame size over the existing levels.

Dr. Bob Kropp from Oklahoma State discussed, "What is Skeletal Correctness." Dr. Kropp pointed out that in selecting straight-lined show animals we can get cattle that lack the needed flex and give to the hock, and the angle and cushion to the shoulder needed to keep them sound. Next on the program was Dr. A.M. Sorrenson from Texas A & M, with the topic of "Femininity and Masculinity." He emphasized that cows must be feminine in their appearance, but bulls should be masculine and we are wrong to select bulls that are extremely late maturing and lack masculinity in their appearance.

The afternoon program featured "The Genetics of Black and Red—Horned and Polled," by Dr. Dick Frahm; "Identifying Genetic Defects," by Dr. Horst Lipold; "Udder Soundness in Beef Cattle," by Dr. Milton Wells; and the program was concluded with Dr. Jim Wiltbanks' presentation on "Managing Beef Cows for Maximum Production."

The second day of the program was held at the OSU Livestock Arena where live animals could be involved with the program. Dr. Robert Totusek, Chairman of the OSU Department of Animal Science, opened the program with his presentation on "The Importance of Visual Evaluation and the Influence of the Show Ring." Dr. Totusek pointed out that most cattle are sold on a visual basis and many of the production traits we select for that have high inheritabilities can be measured visually. He also said the show ring was a great stimulus in changing the type of cattle. The show ring is not infallible and does make mistakes, but it does have a great

In the 1985 National Western Fed Beef Carcass Contest a pen of Limousin bred by Pompadour Hills, Highmore, South Dakota captured the grand championship award. Harold and Marilyn Rinehart are Founder members number 90. Marilyn led the founding of the junior association.

deal of influence on what we as purebred breeders are raising. He pointed out there does not have to be any antagonism between show ring type and the cattle that are the best suited for commercial production, that we can select cattle in the show ring that are the kind of cattle that are profitable to all segments of the industry. Dr. Totusek concluded by pointing out although visual appraisal is very important in and out of the show ring, a wise breeder would not limit himself to visual evaluation only. Breeders should use all the tools available and mix visual appraisal with true performance testing programs to produce the most desirable product.

The first class of live animals were brought out and the audience, as well as the official judges, studied the cattle and marked their placings. The officials were made up of several of the scientists who had been on the program the last several days, renowned show judges and Limousin breeders, as well as some other beef cattle breeders from across the country. They were divided into four different panels and each panel was to discuss the judging classes from their perspective only. The panels were muscling, frame, skeletal correctness, femininity and masculinity.

Between judging class one and two, Dr. Frank Baker addressed the crowd on the importance of performance testing. He said the profit-conscious commercial cattlemen must rely on the purebred industry to performance test their cattle and to identify the strengths and the weakness of the available genetics so the commercial men can use them in their programs. Dr. Don Good from the Kansas State Department of Animal Science addressed the group on the issue of "Show Winners Can Be Useful, Productive Cattle." He pointed out many examples of show winners that had been productive, useful cattle and indicated cattle now winning the shows were closer in type to the profitable cattle in commercial operations than ever before. He cautioned us that it is our general trend to go to the extremes and we may go too far in the direction of frame size without enough thickness. He stated that beauty or eye appeal in the show ring is of no real virtue if performance and usefulness is not part of the same animal.

Three individual classes of animals were judged by both the audience and the committees. In many cases, the committees did not place the cattle the same, as each one was concentrating on a different area rather than trying to put a combination placing on the class. After comments from the muscling, frame, skeletal and femininity and masculinity committees, questions were fielded from the audience. After much discussion and a very informative question and answer period, a combined placing was arrived at. Throughout the morning the committees took advantage of the opportunities to point out animals that were too light muscled or too heavy muscled, or too masculine or too feminine, or ones that did not have enough frame or even one particular cow that appeared to have too much frame. There were also several opportunities to discuss and point out differences in skeletal correctness and freeness of movement.

In closing the live animal evaluation session, I had the opportunity to address the group on the subject of "Different Strokes for Different Folks." The gist of this talk was that individual breeders need to select for things they specifically think are important in their programs and to excel in specific traits which they wish to contribute to the purebred industry. For example, one breeder might select heavily for milk production knowing that his herd will excel in this area and that other purebred breeders can come to him if they need to increase their milk production.

Since the first Limousin were introduced into North America, Limousin steers have earned top billing not only on the rail and in the feedlots, but in the show ring as well. This three-quarter blood Eclair son (right) earned grand champion steer honors for Sonya Deatherage of Stanton, Texas at the Fort Worth Fat Stock Show in 1979. He was bred by the Noffsinger Ranches of Walden, Colorado. In 1979 alone, Limousin dominated the toughest steer circuit in the U.S., also taking reserve grand champion honors at Fort Worth; grand champion honors at San Antonio; grand and reserve grand champion honors at Amarillo, San Angelo and Abilene; grand champion honors at El Paso and at the prestigious Houston Livestock Show. The Houston winner set a record price at that time of $70,000. A year later, a Limousin came back to win Houston and again set another record price of $110,000.

The 1980 grand champion steer at the San Antonio Show (left) was a three-quarter blood Black Jack son shown by Susan Holcomb of Stanton, Texas.

I was also able to reiterate the fact that Limousin cattle are truly different and there is no reason for us to make our cattle look like the other already popular breeds in this country, that we must maintain the true strengths of the Limousin breed and not compromise those strengths in order to improve on some of our weaknesses.

The "Crossroads '79" symposium ended up having a tremendous impact on the direction of the Limousin breed. Limousin breeders and academic people across the country began to realize that Limousin cattle were not supposed to look exactly like the other breeds they were used to seeing. People were convinced it was okay for Limousin cattle to be thicker and heavier muscled than the English breeds they were being compared to. I believe the philosophy was developed in the national board that it was important we maintain our strength, which is leanness, muscling and carcass cutability so we would have something to offer the commercial producers they could not get from the other breeds. The breeders also realized it was much more important to develop structural correctness and pattern than to select only for frame size. For several years following "Crossroads '79" emphasis was placed on levelness of rump, correctness of shoulder, flex to the hock and freeness of movement rather than trying to get the cattle taller.

The Limousin breed was at a crossroads in 1979. I truly believe the symposium helped the breeders select the right direction for the coming years. We were able to maintain the strengths of the breed, improve on several areas that needed attention and yet develop a true identity that set Limousin off as being unique and different and having something to contribute to the beef cattle industry that the other breeds did not excel in. It would have been very easy for the Limousin breed in 1979 to simply follow the other breeds selecting for narrow or flatter made cattle with less muscling and increased frame size in the attempt to get larger, growthier cattle that were more feminine and more fertile. This would have been the wrong direction for Limousin to go, as we would not have maintained our unique identity and the strengths that have made Limousin the "carcass breed" today.

Three years after "Crossroads '79," the board of directors determined it was time for another breeders' symposium. This time it was felt the attention needed to be on performance records and performance programs. The breed had moved along nicely since 1979, but our breeders were not really tuned in to all the tools available to help them do a better job of selecting high performing, profitable cattle. "Direction '82" was held in conjunction with the National Junior Heifer Show in Wichita, Kansas. The current movement in performance programs was towards a total performance system rather than selecting only for heavy weights at weaning and yearling time. Many of the commercial breeders throughout the country, as well as most purebred breeders, had a negative feeling about performance testing. The standard image was simply feeding animals to get them fat so they would have heavy

Jim Davidson, Absarokee, Montana served on the NALF board from 1980-1985. James Dyer, Davis, Texas is serving his second board term. Davidson was president in 1984-85.

weights at weaning and yearling time. The Beef Improvement Federation had worked diligently to develop a total performance program that balanced the traits and put emphasis on all areas of production. It was the combination of moderate birth weights, good weaning and yearling weights, reproductive soundness and efficiencies and using Sire Summary EPD's to develop cattle that were predictable and consistent.

"**D**irection '82" did not have the impact on the breed that we had seen with "Crossroads '79." However, it did set the stage for a new understanding and a new importance of total performance records. Breeders went away from "Direction '82" realizing a complete record keeping performance program was a necessary part of every good purebred operation and that they must learn to use all the performance testing tools available to develop superior, predictable seedstock for their commercial customers.

The show ring has obviously had a large influence on Limousin breeders and the direction of the Limousin breed since the cattle were first introduced in 1969. In the late 60's and early 70's, the entire beef cattle industry was striving to produce a larger framed, leaner looking animal. When the Limousin cattle first came along in the early 70's, they dominated the live steer shows, as well as the carcass shows throughout the country. The cattle had the extra thickness, muscling and tremendous length of body that made them very popular in the steer shows. Limousin cattle first started to show in the early 70's and were being selected along the same lines as the other, more well established breeds with the emphasis on increased frame size, flatter muscling and more femininity.

After "Crossroads '79," the Limousin shows began to change. In breeding shows across the country, judges were starting to select Limousin cattle with slightly more thickness and volume than what they would take in the other breeds. The breeders were taking cattle to the shows that had more natural thickness and muscling and yet were still correct, big volumed cattle and therefore the judges had more of this kind to pick from and use in their championship line-ups. A great deal of emphasis was placed on the correctness of the feet and legs in all Limousin shows. Early Limousin cattle were criticized for being round-butted and sickle hocked. Pressure to correct those two areas was extensive. In no time at all, Limousin cattle were level, square-rumped cattle with straight hind legs. We were still able to maintain the natural thickness and dimension to the quarter, as well as having adequate flex and give to the hock for structural soundness.

Although the general show ring standard was to pick larger framed cattle that were taller, this was not necessarily the rule with Limousin shows. Again, the emphasis was on structural correctness rather than picking tall cattle that were incorrect or bad

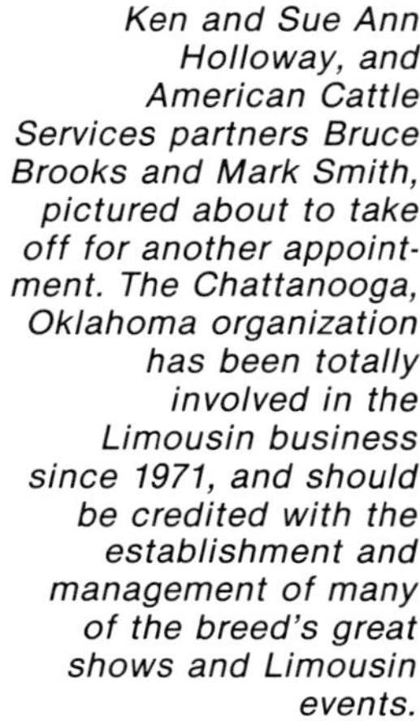
Ken and Sue Ann Holloway, and American Cattle Services partners Bruce Brooks and Mark Smith, pictured about to take off for another appointment. The Chattanooga, Oklahoma organization has been totally involved in the Limousin business since 1971, and should be credited with the establishment and management of many of the breed's great shows and Limousin events.

on their feet and legs. As we look back at this now it was probably a blessing in disguise. We have had several generations of selection for correctness rather than selection for frame and scale. Judges also realized that as they took Limousin cattle with more thickness and muscle dimension through the rear quarter, they had to accept cattle that had more bulk and volume to the shoulder and chest area. There was a time when cattle were being selected that were too shallow-bodied and too tight in their heartgirth and lacked the adequate capacity that was desirable for economic production. This general trend happened in Limousin shows as well as in all other breed shows because those cattle that were shallow-bodied appeared to have more height and frame than the cattle that were deeper bodied.

The shows moved towards larger framed animals with more side view profile and less emphasis on thickness. This was particularly true in the steer shows across the country as steers that were 58 to 60 inches tall were regularly winning the steer shows. If these cattle were weighing less than 1,300 pounds, most of them were relatively narrow and shallow-bodied and did not have the thickness or muscling necessary to produce a very desirable carcass.

Oklahoma State University came back into the picture again in June of 1982 when they sponsored the National Steer Symposium. After two days of presentations based on research information and reactions from all segments of the industry, it was unanimously agreed by those in attendance that it was possible for a grand champion steer to fit the industry. If the steer was going to fit the industry there were some limitations on the carcass weight of somewhere between 670 and 800 pounds. This gave us a live weight range of between 1075 and 1250 pounds. It was also agreed that cattle needed to be thick, heavy-muscled cattle with high cutability, preferably yield grade 2's, and that it would be desirable for these cattle to hang a U.S.D.A. Choice carcass. Based on this discussion, it was concluded that a steer of this carcass weight and quality would be somewhere in the area of 54 to 56 inches tall and would have .3 inches to .4 inches of back fat. This was definitely a real change in the direction that the steer shows had been taking.

In the next couple of years following this National Steer Symposium, the frame size of winning steers came down dramatically. Winning steers ranged from 1,150 up to 1,300 pounds and very few were over 57 inches tall with most of the winners somewhere in the 55-56 inches range. The judges also began to take cattle with more natural thickness and muscling. There was much more emphasis on cattle having large ribeyes and showing the thickness and muscling necessary to produce a high yielding, lean carcass. This whole trend toward smaller framed steers with more natural muscling was a real boost for the Limousin breed. Limousin steers definitely had the muscle thickness that was termed desirable by the industry and also had the eye-appeal and the style and balance that made steer shows attractive to all segments of the industry.

While the steer shows were coming down in frame size, there was not much immediate change in the direction of the commercial cow herds in the coun-

Gene McKown, Norman, Oklahoma, recipient of the 1986 National Western Premier Breeder award, presents the 1987 Premier Breeder trophy to Floyd McGown, New Braunfels, Texas. The McKown and TEXS show strings have been two of the breed's most accomplished exhibitors.

try. The academic people had been promoting to the commercial industry the need for a larger framed, heavier milking female in order to increase their profitability. Most everyone realized that as we increased the size of the cow herd, both from a standpoint of frame and weight, and increased the amount of milk she could give genetically, we created a beef cow factory that could produce a larger calf at weaning time. Most of the commercial cow/calf people were selling calves at weaning time and the total pounds of weight they could sell obviously increased the total number of dollars of income.

Late in 1983 and early in 1984, several scientists began to look at the actual economics of commercial beef cattle production more closely. The watch word of the industry was optimum production versus maximum production. The theory was that not necessarily is more weight better if the cost of getting that weight offsets any added value. One of the very simple examples was additional growth rate and how calving difficulties come along with it. Selecting bulls with high yearling weights, but also with very high birth weights, could create additional death loss at calving time and therefore reduce the total pounds the commercial man had to sell at weaning time. Even though the average weight per calf might be excellent, if the death loss at birth was very high, the bottom line meant lower profitability.

The animal science departments at Michigan State University and at Colorado State University decided to go together and develop a beef cow efficiency forum to look more closely at the efficiencies involved in beef cattle production. In summarizing the conference, Dr. Bob Totusek stated, "The following formula tells the whole story: Profit = Weaning Weight × Percent Calf Crop × Selling Price Per Pound of Calf × Pound of Cows − the Annual Cost of the Cow/Calf Operation." The bottom line was quite simply profit and not maximum production. He pointed out that anything done to increase weaning weight must cost less than the value of the additional weight to increase profit. The same can be said for each one of the variables in the formula. The name of the old ball game was to maximize production. The name of the new ball game is to maximize profits, therefore, optimum production.

The Yackley family has compiled an impressive list of Limousin breeding accomplishments. Probably most respected is their string of five championships for both the pen of bulls and carload of bulls in the National Western yards. Pictured are Bob and son Steve, Onida, South Dakota.

The Triple Crown trophy is awarded to the exhibitor that wins consecutive grand championships at the All American Futurity, the Kansas City American Royal and the National Western. In 1986, the first year of the contest, Spitz Farms, Norman, Oklahoma won both trophies with Spitz Navajo and Spitz Special Effort. Fred and Janell Spitz, owners, are pictured above.

The bottom line of the "Beef Cow Efficiency Forum" was that we are better off with medium to small framed cows that are medium in milk production rather than large framed cows with extremely heavy milk production. Many other people in the academic community interpreted the forum in much the same manner. When all areas of cost were included and all factors of profit were analyzed, there was a great deal of information indicating that more is not necessarily better in terms of larger, bigger framed cattle and heavier milking cattle. It was pointed out there will be some variation in the size of the cow and the level of milk production an individual commercial breeder could support based on the management and the nutrition he has available in his operation. The symposium was really positive for the Limousin breed. We are not one of the breeds with the extremely large framed cows, nor are we a breed noted for extremely heavy milk production.

Another interesting observation that has happened since the cow efficiency symposium in 1984 is more emphasis on carcass characteristics and on beef, the final product. As more and more work has been done by the beef cattle industry to promote our product and to do a better job of convincing consumers to buy beef, we have also looked more closely at what the consumer wants in terms of a beef product. The industry has heavier discounts on yield grade 4's and 5's than ever before and it was particularly true when we had an excess supply of beef in the summer of 1985. For a breed that excels in leanness and carcass characteristics, it is very rewarding to see emphasis placed on the final product. In years past, most of the efficiencies of beef cattle production dealt around weaning weights and pounds of calf rather than the total value of the carcass produced. We all realize that the final product, beef, is probably seven or eight steps away from the purebred industry and therefore it is much harder to get paid a premium for a superior carcass than it is for increased weaning weights. The Limousin breed has continued to prove its superiority as the carcass leader and we are convincing people across

A most familiar shot in the Limousin scene in North America— Jimmy Linthicum in the show ring. The Linthicum family including dad, sons Jimmy, Timmy and Teddy and daughter Cheryl have been consistent winners in open and junior competition. Cheryl served as National Limousin queen and as president of NALJA.

> **Limousin Tops Them All**
>
> The National Limousin Sale not only set a new record average this year at $12,090, it also produced the highest average of any breed sale in Denver this year. The table below charts the performance of the National Limousin Sale since its inception 17 years ago.
>
> **National Limousin Sale**
>
> | Year | # Lots | $ Gross | $ Average |
> |------|--------|---------|-----------|
> | 1971 | 103 | 148,114 | 1,438 |
> | 1972 | 80 | 227,330 | 2,839 |
> | 1973 | 53 | 229,060 | 4,320 |
> | 1974 | 69 | 468,900 | 6,796 |
> | 1975 | 84 | 310,975 | 3,702 |
> | 1976 | 57 | 194,350 | 3,410 |
> | 1977 | 54 | 152,125 | 2,817 |
> | 1978 | 57.5 | 223,575 | 3,888 |
> | 1979 | 45.5 | 237,900 | 5,273 |
> | 1980 | 52 | 370,115 | 7,118 |
> | 1981 | 65 | 322,275 | 4,958 |
> | 1982 | 56.5 | 560,570 | 9,921 |
> | 1983 | 54 | 444,300 | 8,228 |
> | 1984 | 48.9 | 550,140 | 11,250 |
> | 1985 | 67.7 | 786,350 | 11,615 |
> | 1986 | 50.9 | 493,650 | 9,698 |
> | 1987 | 41.5 | 501,750 | 12,090 |
>
> *March 1987/Limousin World*

At the National Western Stock Show the past five years auction sales of purebred and commercial cattle have totaled between 7.5 and 9 million dollars. Attendance at the 10 day January event in 1986 exceeded 460,000.

the country that you can take off the excess fat in a single cross with Limousin.

The other thing that has come out of the shift from maximums to optimums is the need for more predictable genetics. The commercial industry is more in tune to the genetic needs of their individual programs than ever before. A commercial operator knows whether he needs to increase his milk production or needs to have more growth rate or simply needs cows that can utilize a limited amount of poor quality forage.

In the areas of the performance traits we can measure, such as birth weights, growth rate and milk production, the new Expected Progeny Differences are going to be very significant. They are the best tool ever developed by the beef cattle industry to determine the predictability of the genetics we have to sell. As we lean more toward the optimum theories of production rather than maximums, it becomes not as important how much a bull can change in trait, but rather are we truly sure the bull can make this change in his progeny? The new *Sire Summaries* and the current individual animal EPD's are the best tools we have to provide predictable genetics to the commercial cattle industry. If a commercial cattleman wants lighter birth weights, then we can use EPD's to provide him with bulls that will give him the genetics for smaller birth weights in his program.

The direction for the breed in the near future seems to be relatively stable. All indications are that in years to come we will continue to refine the theory of optimum levels of production rather than maximums. Expected Progeny Differences will become more and more accurate and we will be able to evaluate them and calculate them on younger animals. This should allow us not necessarily to provide the commercial man with our interpretation of superior genetics, but to provide him with the predictable genetics to make the changes in the genetic base of his program that he feels are necessary.

The first national Brahmousin show classification was set for the 1985 National Junior Heifer Show. Kristi Kent, Sweetwater, Texas led Magnums Molly to the championship. The daughter of WRC Magnum 266N was bred by Bob and Vivian Good, Dilley, Texas.

Chapter 4
Breed Promotion

In going back and looking at breed promotion activities for the last nine years, I have decided to divide this chapter into three main sections, junior programs, general promotional activities and advertising. In discussing the junior programs, I don't think anyone would argue the fact that it is the real basis for long-term breed growth and expansion. The junior members who are involved in breed activities today will be the members of the association and the strength of our breed in future years. We've also found out over the years that many times the children get involved with Limousin cattle first and eventually they get their parents enthused and they become involved also.

The first official National Junior Heifer Show was held in 1976 in Des Moines, Iowa. It was part of the World Futurity and was not really planned as something that would continue to be a national annual event. The success of the Iowa Junior Show created a desire and interest to have it on an annual basis. In 1977, the National Junior Show was held in Springfield, Missouri at the Ozark Empire Fair Grounds. Again, this event created a great deal of enthusiasm and excitement in both the young people involved and in their parents, and it established

a real pattern for the junior shows. There has been a junior show every year since and they are now being planned and scheduled two and three years into the future.

The National Junior Show now includes many more activities than simply showing heifers. Besides the outstanding quality of cattle exhibited during the show itself, juniors are given the opportunity to compete in several activities. There is a showmanship and fitting contest, a team fitting contest by state, speech contests with four different divisions, quiz bowls, educational programs and judging contests. This show has been designed to provide several days of activities where kids get to know each other and learn to work together.

Several years ago, the junior bylaws were rewritten so the association is now under the leadership of a 10-member junior board of directors. Their directors are selected by a committee (appointed by NALF) based on a written application submitted to the national office. Out of the 10 junior directors, the junior membership elects its officers each year at the annual meeting held in conjunction with the National Junior Heifer Show. They handle the weighing of cattle and measuring, checking of tattoos, organization of all events, as well as handling the PA system during the show. By letting the juniors be totally involved in controlling the success or failure of their show, they have learned a great deal about what is necessary to manage and pull off events of this kind. I am sure as they get older and become involved in these events as adults, they will find their experience as juniors to be very helpful.

There are also a couple of special programs available through the Junior Association. One is the new Awards of Excellence program which is designed as a financial award to junior members that have exhibited excellence, not only with Limousin cattle, but in other areas of agriculture as well. This new award is sponsored by the interest earned from a large donation that was received a few years ago during the National Western Stock Show. The other new program is the State Field Day Fund. This is another cash award given to state associations to help put on state junior field days each year. The funds for this particular event also come in the form of interest from a CD that was a large cash donation. The National Junior Program is supported each year by numerous donations from across the country. NALF has currently been donating $10 per each junior animal registered and $10 for each new junior member each year to the Junior Fund. This amounts to about $13,000-20,000 per year. The rest of the money needed to put on the events during the summer and sponsor other activities is somewhere in the area of $30,000-40,000 per year. This additional money has been raised each year through various fund-raising activities and through the support of numerous breeders across the country. A very special thanks needs to go to all those breeders and Limousin supporters throughout the country who have helped our junior program throughout the years.

General promotional activities cover a wide variety of things the Foundation has been involved in over the years. One of the most successful is NALF's program to provide display booths to all our state associations for their activities or events. Currently, we have six portable display booths that are being shipped across the country. NALF pays the freight on these, both to the state association or the breeder and also pays the freight back to the shipping depot here in Denver. The displays are very attractive and are designed to promote lean Limousin cattle with the slogan "Trim Off The Excess Fat In One Cross With Limousin." The Foundation has been very active in having displays at the National Cattlemen's Association Convention and The Future Farmers of America Career Convention each year.

The National Association provides state and local associations and their members with an attractive, professionally designed booth for their use upon request. One of the booths is pictured manned by the juniors from the South Eastern Limousin Breeders Association.

Dan Wedman, Yukon, Oklahoma is owner publisher of the Limousin World, *the official publication of the North American Limousin Foundation. He is on the board of the Livestock Publications Council. He started working with the breed as a fieldman after graduating from Oklahoma State University in 1974.*

These two large events give us an opportunity to talk with and visit a large number of commercial and purebred producers at the NCA Convention and a chance to expose ourselves to the most outstanding young FFA people in the country during their convention in Kansas City.

The Foundation currently has a film strip entitled "Seven Thousands Years New," which is a 15-minute general promotional film about Limousin. This film is currently shown in approximately 150 schools and meetings each month during the school year and last year reached over 40,000 viewers. Primarily, it is used by vocational agriculture teachers and 4-H groups. The film was developed in about 1975 and is quickly becoming outdated. The Foundation has recently completed a video tape "Lean on Limousin". It points out the needs of the new lean beef movement and how Limousin fits in today.

Each year the Foundation prints large quantities of fliers and brochures to be handed out at meetings and fairs across the country. The fliers are primarily designed to promote and explain the advantages of Limousin cattle in commercial beef production today. We also provide balloons and other free material that can be given away, and posters that can be used as decorations anywhere Limousin cattle or people are gathering.

Since early in the 1970's, the Foundation has been involved in providing subscriptions to our official publication, to university professors and to extension people that are involved with beef cattle. This has been a tremendous help in keeping people in the academic area informed about what is happening within the Limousin breed. We have been able to provide these subscriptions to the academic area through a joint sponsorship with our official magazine.

The Foundation and Limousin breed in general has always been blessed with having an excellent quality breed publication. From the early days in 1970, the *International Limousin Journal* was one of the most outstanding publications in the country. It did a great deal to encourage more and more people to become involved with Limousin cattle and the Limousin breed. The magazine's founder, owner and publisher until 1980, Dale Runnion, is known to be one of the best there is in the world today in the field of livestock journalism and particularly, in purebred beef cattle publications. After a short period of time, in the early '80s, when Dale was not involved with the *Limousin Journal,* he came back in the fall of 1983 and started the *Limousin World.* This new magazine was started because of problems that had been created with the old *Limousin Journal* after Dale left. The *Limousin World* is now our official publication and, again, is providing us with an outstanding, high-quality magazine under the guidance and direction of owner Dan Wedman and former owner Dale Runnion.

Livestock shows have a great deal of promotional value in the industry here in the United States. The enthusiasm and excitement that shows create, as people prepare their cattle for head-to-head competition, is unequalled by anything else in the livestock industry. In the early 1970's the Limousin breed received a great deal of visibility through steer shows in which Limousin cattle competed. For several years, the majority of the steer shows were won by Limousin cattle, either in the live animal division or in the carcass division. These steer shows gave the

The American Brahmousin Council set up this attractive display of Brahmousin live cattle in the yards at the 1986 National Western Stock Show in Denver. The display illustrated the percentages necessary for purebred Brahmousin registration.

breed a tremendous opportunity to display to the industry the kind and type of cattle we were capable of producing with Limousin genetics. As the breed grew and developed, so did the showing of registered Limousin cattle. Not only does the show ring provide an excellent opportunity for breeders from different parts of the country to compare their cattle on the same turf, but it provides a very positive social atmosphere in which breeders and other interested people can get together and discuss Limousin and the cattle industry in general.

The National Western Stock Show has provided a tremendous opportunity for the promotion of Limousin cattle by having our national show and sale here the last few years. With the Foundation office currently adjacent to the stock show grounds, we have a distinct advantage in promoting Limousin cattle throughout National Western week. It also has become a very important status symbol within the breed and industry to have one of the champions at the great National Western Stock Show. The other exciting thing about the National Western, and maybe even more practical, is the Pen and Carload Bull Show that is held in the yards. This is one of my favorite events, where groups of three and carloads of eight or 10 are displayed running loose in the judging arena. The basis of any purebred cattle organization is your ability to sell bulls to commercial cattlemen. Seeing these bulls shown loose in groups is without a doubt one of the most impressive and exciting events we have in the Limousin breed. For several years, we have ranked second behind the Hereford breed in the total number of bulls shown in the yards and for the last two years Limousin bulls sold at private treaty in the yards have been the high-selling breed when compared with all breeds that sold ten head or more.

Advertising has always been a difficult thing to analyze and discuss. I think we all realize advertising is an important part of any business, but is also the part most difficult in which to measure direct results. Based on the Board's policy, our advertising has been directed toward selling bulls to com-

Doug Harris represents the Limousin World *in the southeastern United States and eastern Canada. Pictured right is Randy Bollum, a Blue Earth, Minnesota native and the owner publisher of the* Limousin Leader, *Calgary, Alberta, Canada.*

Sale day at Worrell Land and Cattle Company is an event. Gene and Anne Worrell and their record selling production sales have drawn breeders to the farm's historic site near Charlottesville, Virginia in record number.

mercial cattlemen throughout the country. Early in 1978, Dale Runnion and myself developed a couple of ads around the theme "The Word is Limousin." The ads used the new information we currently had available in calving ease and profit per cow that came out of the OSU report. The advertisements were extremely successful in developing a knowledge throughout the industry that Limousin was a breed of cattle and some of our strengths were calving ease and profit per cow. We also developed a four-color, 24-page brochure called "The Word," with the help of Bob Crook, who was at the time the editor of the *International Limousin Journal.*

Our next major advertising campaign was also developed with the help of Mr. Runnion. The theme was "Discover the Extra in Limousin Cattle." The word "extra' appeared quite large on all the ads and was the teaser to get people to read the ads more closely and to discover the extra in Limousin. These ads had a new look in that rather than using line art as "The Word" ads did, they used large pictures. Most of the pictures were black and white but we did begin to use added color where available and ran a few ads in four-color. There were basically three different ads in the "extra" theme, one to promote feeder cattle, one to promote bulls and one to promote females.

Our next advertising campaign was more general and has been keying on the word "Limousin" across the bottom so they could be recognized from one publication to another. These ads with the distinctive border also helped us to differentiate our ads when we were running partial page ads in newspapers. These ads were developed in-house by Jill Wood, our Director of Communications. Most recently, the ads have been taking on a more general appearance, promoting the leanness of our cattle,

the fact that Limousin has been a leader in the performance area and the productivity of the Limousin females. It has always been the philosophy of our Board that we should be positive in our advertising rather than negative. What we are saying here is we promote the positive aspects of our cattle without trying to run down another breed. I think this is very important in that most of our potential customers have cattle of a different breed and would be using Limousin in a rotational crossbreeding system. We must be positive in our advertising and complement other breeds rather than being negative in trying to set ourselves apart.

Current advertising uses the theme "Lean on Limousin". The ads were designed to take advantage of the new trends in the beef industry for a leaner animal with less fat. These ads have been very timely and well accepted by the industry.

In June of 1984, the Board approved a research survey to be conducted by Miller Research to evaluate the effectiveness of our advertising/promotional efforts. The results were very gratifying with nearly 98 percent of all people surveyed indicating they had seen our ads and were aware of Limousin cattle. Our ads ranked consistently in the top two or three breed association ads and were thought to be professional and accurate, including useful information in all cases. This is extremely encouraging when one realizes that of the 12 or 15 other breeds we compete with, at least four and possibly five of those spend more money on advertising each year than we do. Three of the large breed associations spend three to four times as much on paid advertising as the Foundation has currently budgeted. The fact that in many cases our ads are ranking with or above those that spend a great deal more money gave us encouragement that we were on the right track. One has to realize that whether or not people remember your ads does have a great deal to do with how many times those ads are placed and in how many different magazines those ads appear, as well as the design and content of the ad.

Breed promotion is definitely a difficult area to define and measure in terms of success. However, I believe the programs the Board has okayed in the past and the plan we have for the future will, indeed, keep the word "Limousin" in front of the cattle industry and increase our particular share of the market over the years to come. By having well developed advertising that's positive, supported by promotional activities and junior programs that help keep people enthused and excited about the business, we will continue to expand the Limousin influence throughout the cattle industry.

Chapter 5
Leadership

The North American Limousin Foundation has been blessed over the years with an extremely strong Board of Directors. In the nine years I have been associated with the breed, the Board has always been in the position of putting the breed first and their individual interests second. This is extremely important if a breed association is going to make progress and grow and expand in a very competitive market. Everyone has to understand the individual needs of each breeder are somewhat different than what might be in the best interest of the breed in total. It is a very difficult job to serve on a board of directors and it is extremely challenging to sit there and try to make the right decisions for the good of the entire breed rather than what particularly pertains to your individual program. The Board is made up of 15 lifetime members who are currently active in the breed. The Board normally has a fairly wide cross-section, including people with very small herds as well as those with large herds. The Board reflects the people who make 100 percent of their living from the cattle business as well as those who are involved in other areas of agriculture or have other businesses outside agriculture.

Many times a year, we get asked the question, "What is the average member like who does business with NALF." It is very difficult to determine the average member, but we do have some statistics that will be interesting in terms of our membership make up. Last year, 73 percent of the people who did business with NALF recorded 10 or less animals with the Foundation. Of those who had transfers processed by NALF, 81 percent transferred 10 head or less. This definitely points out that NALF is an organization of many small members spread throughout the United States with an interest in Limousin cattle. Only 2.5 percent of our total members recorded over 50 head each during the past year, however, they did account for approximately 27 percent of our total registrations. Obviously, the

At the 1987 Annual Limousin banquet Dale and June Runnion were recognized for their dedication and support of the Limousin breed. The Runnions were presented an Honorary membership by the North American Limousin Junior Association in 1985.

Lewis Bingham, Cecil Johnson, Tommy Stewman. Bingham and Johnson have served as Texas Limousin Association presidents and Stewman is now vice president. Each has actively participated in state and national Limousin events as individuals, with their families and through their respective ranches.

large breeders are important to us in terms of the number of registrations processed, but we are basically an organization of many small members. In the past 12 months, there were 3,645 NALF members who had work processed by the Foundation. If each member would register and transfer two more calves this coming year, the Association would be back near the level of registrations and transfers we had two years ago. Just stop and think, with this many small members out there, our organization has a tremendous amount of strength and staying power.

Since I have been working for the Foundation, it has always been the philosophy of the Board that their position was to make Board policies and establish guidelines and direction, and it was the responsibility of the executive vice president and his staff to carry out the day-to-day activities and to execute the programs and policies the Board establishes. From my standpoint, this has been a very positive and enjoyable relationship between the staff and the Board and allows the staff the flexibility to be more productive and efficient in achieving the goals the Board has set. The past board presidents who are still actively involved in the breed have been used for several years as the nominating committee. This group of men have done an outstanding job in finding qualified, dedicated people to serve on the Board each year. With a 15-member Board and three-year terms there is a possibility for five new members each year. As a general rule, we normally end up with two or three new Board members every year, keeping new ideas and new input coming into the Board at all times.

Every year at the annual meeting, the Board of Directors has conducted several open forums and open committee meetings to get input from the membership and to create new ideas. The committees meet in an open session where anyone can make suggestions or discuss problems or issues they think should be addressed by the Board. The committee can then make a recommendation to the Board of Directors, which will be acted upon by the Board at their next formal meeting. The Board has always felt it is an association of members, for the members, and they should do everything possible to be responsive to the needs and desires of the membership. It

George Cocoros, Gaffney, South Carolina receives a trophy from South Eastern Limousin Breeders Association (SELBA) president Bob Dwyer, Monroe, Georgia in recognition for his efforts in behalf of SELBA. Cocoros is NALF Founder member number 63.

should be pointed out at this time, the membership is a very diverse group and the Board must balance all their decisions by looking at the total composition of the membership rather than only at a small group of members from one particular area or with one particular interest.

At this time in our history, the Foundation is fortunate to have a strong financial position with adequate cash revenues. The Foundation was started in 1968 on a sound financial basis by requiring each of the founding members to put up $2,500. This turned out be a very important decision as it provided up-front funds for the Foundation to become established and to get involved in importing Limousin cattle and semen for the founder members. Over the years, the Foundation has been able to build cash reserves to a point where we recently have reserves that are equal to one year's operating expenses. This is normally considered to be adequate cash reserves for non-profit organizations such as ours. In addition to the cash reserves that have been accumulated, the Foundation has invested more than $1,000,000 in developing one of the most up-to-date and complete computer record systems anywhere in the nation. A great deal of money was spent developing software that can provide information to our members as quickly and efficiently as possible. Nearly every piece of information we have, in regard to the Limousin cattle that have been registered, is immediately on-line. On-line means that the information can be accessed by a computer terminal in a matter of seconds. This allows our staff to get information quickly and efficiently for our members, providing answers to nearly any question a member might have while they are still on the telephone.

The Foundation has been very lucky in that our membership has supported us with registration and transfer fees of both bulls and females. Most people do realize that registrations and transfers are our main source of income. It is very important that our members register and transfer the bulls they sell to commercial cattlemen. The registration certificate and the performance record that goes with it provides the new commercial buyer with all the information available on the bull or bulls he has purchased. By transferring the animals to him, he gets the official documentation from our office and also

Ron and Carolyn Holland, Osage City, Kansas have been two of the breed's hardest workers for many years. Both have served as officers in their state association. Ron is a member of the national board and Carolyn has served as an officer of the National Limouselles several years.

becomes an associate member of our organization. From time to time, we are able to send direct mailings to those people who have been past customers and who have had animals transferred to them in order to continue to promote the breed and also to find out more about how the cattle are working for them.

We have just recently finished a questionnaire that was mailed to over 14,000 associate members who have purchased bulls in the last three years. The results of the questionnaire were very gratifying and proved that the Limousin bulls are working extremely well for our commercial bull customers. In several different traits and categories, the bull buyers were asked to respond as to whether Limousin rated excellent, above average, average, below average or poor. Eighty percent or more said that Limousin-sired calves were above average or excellent in calving ease, calf vigor, and growth rate. When looking at the carcass traits, 88 percent or more of those surveyed said that Limousin calves were above average or excellent in feed efficiency, average daily gain, dressing percentage and acceptable carcass weight. When we looked at the traits of quality grade, 84 percent said Limousin were above average or excellent, and when asked about packer acceptability of Limousin cattle, 80 percent said that the cattle were above average or excellent.

Ron and Tony Simek, Little Mondeaux Limousin, Genoa, Nevada have an appetite for quality Limousin and competition that seems unquenchable. Famous cows like Pertinente, Crystal and Miss Snowflake are owned by the Simek firm. Their Tombstone Pizza and beer "for all" in the Denver yards during the carload and pen shows were boosters for the event and breed.

doing a good job in many areas that are important to the commercial cattlemen, but we do need to work on some of our weaknesses. If we intend to survive in the beef cattle industry, we must respond to the needs and wants of our commercial cattlemen and provide them with the kind of Limousin bulls that will help them improve their cow herd and make their program more profitable.

Our Foundation is in a strong financial position. In 1985 and 1986 we experienced some decrease in registrations and transfers due to the economic difficulties of the cattle business in general. As a general rule, the decreases we have experienced have been less on a percentage basis than most of the other beef breed associations. The planning and foresight on behalf of the directors to build a strong cash reserve will allow us to go through some of these difficult times without reducing the services or programs we have available for our membership. We appear to be back on track in 1987. Registrations and transfers are up considerably as compared to the previous two years.

When our commercial bull buyers were asked to rank the Limousin replacement females they had put back in their herd, the number one ranking trait was the growth rate of their calves. Of those producers responding, 90 percent said Limousin females were above average or excellent in the growth rate of their calves, and this was followed by 75 percent ranking udder soundness and calving ease as above average or excellent. The traits of fertility, milking ability, longevity and flushing ability were in the range of 56 to 60 percent of the producers rating them above average or excellent. We did find that between 2 and 11 percent of those surveyed felt that the Limousin females needed some improvement in the above traits.

Without a doubt, this kind of information gives us some real guidelines on the important traits that we need to be selecting for. It shows us that we are

Andy Rest joined the Limousin Journal *in 1977 and later the* Limousin World. *His influence and assistance to breeders in the wide area of the Western United States and Canada he travels is appreciated. Wes Ishmael, editor of the* Limousin World *is pictured right. Both are graduates of Colorado State University.*

National Carcass and Feedlot Evaluation Program

The program was approved by the NALF board to gather current data on the breed and to provide breeders with an opportunity to compare their cattle in terms of carcass merit. Phase I of the test reported the following averages:

ADG—2.8 lb./day; **WDA**—2.59 lb./day; **Carcass weight**—807 lb.; **Ribeye Area**—15.04 sq. in.; **Yield Grade**—1.85; **Quality Grade**—high good; **Carcass Value per Day of Age**—$1.47 compared to current industry average of $1.25.

In the summer of 1985, the Board made a commitment to take the time and spend the effort necessary to develop the first association long-term strategic plan. The philosophy behind the development of this plan was to give the Board and the staff some direction, not only for the next year, but for the next five years. It was also felt the development and the publishing of a five year plan would give the membership an idea of the direction the Board intended for the Foundation and some idea as to what our priorities were, both short and long term.

The beginning of a formal long-term plan is to do a situation analysis. The Board did this in the summer of 1985. They took an in-depth look at the total cattle industry in general, the purebred beef cattle industry and then, particularly, the Limousin breed and how it relates to the total purebred picture and to the total cattle industry. Part of the situation analysis was to look at the strengths and weaknesses of each segment of the industry, as well as the opportunities and the threats that might prevail. The Board agreed the number one strength

of the Limousin breed was our cutability, or our leanness. The fact that we have the ability to change cattle that genetically are yield grade 4's into yield grade 2's with one cross of a Limousin bull puts us in a very enviable position within the industry. Other strengths identified with the carcass or feedlot industry were feed conversion, dressing percentage, loin eye size and a very highly palatable product with less fat. Other strengths included calving ease, calf vigor, middle-of-the-road size, moderate milk production, structural soundness, aggressiveness of bulls, low maintenance feed costs and longevity. Additional strengths included our performance programs that identifies our genetic material, good cattle people as a base with a mix of businessmen and people outside of agriculture and a breed that is well known throughout the industry. Several weaknesses were also identified that need to be worked on in the future. Many of these areas have seen a great deal of improvement already, but the breeders still must concentrate on these in order for us to produce a better product. The weaknesses identified were: testicle size, late puberty in females, disposition, the old image of Limousin that were double-muscled and crooked-legged, and a smaller number of total cattle, making it more difficult to identify

Leonard Wulf, Morris, Minnesota receives a hand shake from 1986 NALF president Gene Raymond, Garnett, Kansas as he assumes the NALF leadership role for 1987.

The 1987 NALF Board includes: (seated L to R) Officers, Greg Martin, Executive Vice President; Ron Holland, Member at Large, Osage City, Kansas; Bruce Waddle, Vice President, Platteville, Colorado; Leonard Wulf, President, Morris, Minnesota; Tom Grapner, Secretary, Barnesville, Georgia; Bob Coscia, Treasurer, Springfield, Missouri; Gene Raymond, Ex-Officio, Garnett, Kansas. Directors (standing L to R) include: Jerry Bradley, Braman, Oklahoma; Jim Fawley, Lynchburg, Ohio; Pete Parker, Weed, California; Rolf Mosbo, Rembrandt, Iowa; Carl Crabtree, Grangeville, Idaho; Bill Bridges, Miami, Oklahoma. Not pictured are James Dyer of Fort Davis, Texas; Vernon Holcomb of Stanton, Texas; and Carl Johnson of Brandon, Florida.

Limousin carcasses all the way through to the consumer. The Board felt the strengths far outnumbered the weaknesses, and we could concentrate and improve our weaknesses while maintaining the strengths the breed already has. After a great deal of thought, the Board developed the following mission statement:

"The mission of NALF is to develop, improve and promote the Limousin breed of cattle while enhancing the profit potential for our members."

Five general objectives were identified under our mission statement.

1. To communicate the merits and to expand the acceptance of Limousin cattle in the beef industry.

2. To provide leadership, direction, information and service to our breeders.

3. To involve all NALF members in the performance records program.

4. To cooperate with all segments of the beef industry to increase the acceptance of beef and the profitability for cattlemen.

5. To assist in the development of the Brahmousin breed.

In developing the above mission statement and the general objectives, the Board felt this would serve as a long-term plan or blueprint to help us and the Foundation achieve our desired end point. The Board realizes in order to reach worthy goals and objectives, everyone must be pulling in the same direction. It is the hope of the Board and the Foundation that the long-range plan will help all of our membership to get behind the Foundation and help us pull in one direction. The following are the specific goals the Board listed under each of the five general objectives. Many of these goals have already been completed and the plan will be updated to include new goals for the 1987 plan.

General Objectives and Specific Goals:

1. To communicate the merits and to expand the acceptance of Limousin cattle in the beef industry.

   1.1 Continue the "Lean on Limousin" Feedlot and Carcass Evaluation program; encourage the development of state association feeder calf sales; gather additional feedlot data and expand our Limousin Feeder Cattle Directory.

   1.2 Increase annual registrations, transfers, and new members to NALF.

   1.3 Expand information about Limousin to university personnel and livestock judges.

   1.4 Place more articles in beef publications on the merits of the Limousin breed.

   1.5 Place more advertising in beef publications directed toward commercial cattlemen.

   1.6 Develop films, video tapes and slide sets for breed promotion.

   1.7 Expand the participation in carcass shows, feeding contests, displays, and open and junior Limousin shows.

2. To provide leadership, direction, information and services to our breeders, and to promote and improve the Limousin breed.

   2.1 Work with our industry advisory committee comprised of commercial cattle operators and relay their concerns and suggestions to our breeders.

   2.2 Conduct a workshop for state association leaders in January of 1987.

   2.3 Encourage ethical standards within our membership and to develop a brochure for potential buyers and sellers.

   2.4 Develop video tapes for members on registration procedures, NALF services, etc.

   2.5 Expand junior activities and educational programs at all levels.

   2.6 Investigate alternative facilities to accommodate future needs of NALF.

   2.7 Initiate new research and disseminate information on Limousin cattle.

   2.8 Explore the expansion of the NALF staff to assist in member services and public relations.

   2.9 Schedule and conduct an educational symposium for our membership in the summer of 1987 in conjunction with the National Junior Heifer Show and the Futurity.

   2.10 Encourage our members to maintain the breed strengths of muscling, trimness, calving ease, fertility and moderate carcass size; and to inform them that selection for excessive frame size can lead to problems with calving difficulties, late maturity and carcass weights above acceptable levels.

   2.11 Encourage our members to select for improved fertility and reproductive efficiency by encouraging and assisting in measuring and selecting for larger testicles in bulls and early puberty in females.

   2.12 Encourage our members to select for good dispositions and to cull animals with unruly dispositions.

   2.13 Encourage our members to use 50% or greater registered Limousin cows for use as embryo transfer recipient cows, due to the inability of the industry to accurately evaluate the performance data of calves that are carried in and nurse recipient mothers of a breed different from their biological parents.

   2.14 Encourage natural production (carrying and nursing) of at least one calf before a female is used as a donor in embryo transfer programs.

3. To involve all NALF members in the performance records program.

   3.1 Maintain and improve the Sire Summary, EPD's (Expected Progeny Differences), and other programs that identify genetic merit and predictability.

   3.2 Provide workshops and promotional material on performance records and programs at state, regional and national levels.

4. To cooperate with all segments of the beef industry to increase the acceptance of beef and the profitability for cattlemen.

Rosalie Smith is serving as the 1987 president of the National Limouselles. Her service to the Limouselle and Missouri state organizations extend over a long period. She and husband Tom breed Limousin near Smithton, Missouri.

4.1 Work with other organizations to develop a marketing structure that recognizes the true value of lean beef carcasses.

4.2 Continue membership in, and support of, industry organizations that are consistent with NALF objectives and goals, such as NCA, USBBC, BIF, BIC, and others.

5. To assist in the development of the Brahmousin breed.

5.1 Cooperate with the American Brahmousin Council in the promotion of Brahmousin.

5.2 Work toward a complete performance records program for Brahmousin.

I personally, and the Board of Directors, are very proud of our strategic plan. The directors spent a great deal of time analyzing the total beef industry, with particular emphasis on the purebred segment and the Limousin breed. They looked at the current situation and the opportunities and threats that confront the breed in the future. As they drew their conclusions, they developed a plan that specified the objectives and goals the Foundation wishes to achieve. The plan will serve as a basis for budgeting and management decisions for the North American Limousin Foundation. Because good corporate plans are dynamic and are changed and updated frequently, the Board will review this plan periodically and make changes and additions as necessary and appropriate. It is my hope you will discover in studying these goals and objectives that the Foundation does have a plan and is following a blueprint for success. You have probably noticed that the plan does not mention the Foundation's day-to-day activities of keeping the Limousin herdbook. This, of course, is the foremost purpose of the Foundation and should be understood as NALF's primary reason to exist. The plan focuses on the new objectives and goals which will carry us forward into the future with confidence and strength as Limousin continues to expand its influence on the total beef cattle industry.

Chapter 6
Conclusion

In concluding my portion of this book, I would like to again point out that these are my thoughts on some of the important happenings, events and issues that have taken place during my reign as executive vice president. I purposely chose not to mention specific breeder's names or specific bloodlines of cattle, as I feel this information has been documented over the years in our official publications. It was my intent more to give you an overview of the evolution of this breed of cattle through the last nine years, from being the new kid on the block to being one of the most important breeds in this country today in terms of the genetic contribution we can make to the profitable production of lean beef. It has been my pleasure to be associated with the Limousin breed for the last nine years. I wish to thank the Board of Directors for their support, encouragement and guidance and also, I wish to thank the membership for the opportunity to have worked with them and served them in my capacity as executive vice president of the North American Limousin Foundation.

Part Four 1968-1987
The Canadian Herd

by Dale F. Runnion

Chapter 1
Introduction

Canada is a country of 3.85 million square miles and some 25.6 million people. The wide fertile prairies of the three western provinces—Alberta, Saskatchewan and Manitoba are painted with fields of waving barley, wheat, oats and brilliant yellow rapeseed.

The broad ranges of native grasses intermingled with improved pasture grasses afford an ideal environment for beef cattle production. Eighty-one percent of the 2.408 million beef cows are in the four western provinces. Another 11% are in Ontario.

Some 1300 miles of scenic and recreation country in western Ontario bounded on the south by Lake Superior and Lake Huron separates the western provinces from the heavily populated countryside and fertile farms of the southeastern part of the province. Dairy, poultry and cattle feeding abound in this cereal producing area close to the heavy population areas of Toronto, Ottawa and Montreal, Canada's largest city.

As would be expected, a bigger part of the fed heifer and steer population shifts to the east with 37.8% of the 1.359 million fed head in Ontario. The province of Quebec lies to the east of Ontario. It has a beef cow herd of 170,000 cows. Alberta with 1.13 million beef cows feed out 30.6% of the cattle in Canada. There are 3.179 thousand beef calves under one year of age not included in the above 1987 estimates.

Like its mother country, Great Britain, the pure-bred breeders of Canada have had great pride in the fact they have been a seedstock nursery for the United States—and the world, for that matter. Clydesdale, Belgian and Percheron horses were bred

A map of Canada showing the provinces and principal cities.

in great numbers in the draft horse days, and are still used in the north where winter snow presents a major problem to feeders. Great seedstock herds of Shorthorns were concentrated in the lower Ontario peninsula—Herefords and Polled Herefords in the west. In Ontario the Edwards brothers and Weldon herds, and the Matthews of Alberta were leading North American Angus breeders. For as long as one can remember eastern Canada has been one of the world's largest exporters of high performance Holstein breeding stock.

To this early and continuing history, Winfields Thoroughbreds and several Standardbred herds in Ontario, including the Armstrong brothers, have set the pace for modern day horse racing in North America.

It is little wonder then that the officials of Canada reacted quickly in 1964 to establish a quarantine station when the demand for Charolais from France became apparent. The need to change the appearance of the beef herds on the North American continent was inherent and breeders were looking to the European breeds new to this hemisphere for this change. Stations were established at Grosse Ile and later at St. Pierre.

Grosse Ile, a Canadian owned island in the St. Lawrence River, was the site of the first quarantine station for bovine imports in Canada. The first Limousin bull, Prince Pompadour, came out of the Grosse Ile station in 1968.

As additional facilities were needed, the French government requested a quarantine station on the French owned soil so the station on St. Pierre was built to handle the overflow of imports. When numbers became so great, a third station was established on the island of Miquelon. It was used for only one importation.

Applications for permits were screened by the government giving preference to serious breeders or future breeders before permits were granted. To further the seedstock producing theory export rules were established on all bovine imports into Canada. It stated simply: All male and female imported cattle that have been in Canada for five years following their release from quarantine will be eligible for export. This regulation was in effect until October 1971 when the time imports must remain in Canada before they were eligible for export was changed to three years. First generation males (of imports in Canada for three years) and females (from imports in Canada for five years) were made eligible for immediate export under the new 1971 ruling.

Chapter 2

Organizing the Association

"**I**t's time we start a Canadian Limousin Association."

The statement lacks a specific author. However, it expressed the unanimous feeling of four Alberta cattle people in January 1969. The four, Neil McKinnon, Bassano; Sherm Ewing, Claresholm; and Byron and Laura Palmer, Midnapore, were visiting after a Limousin meeting in Denver. It was obvious to them that Canada needed a Limousin organization and the time had arrived to get it started.

There appears to be no record of the first meeting other than the memory of those attending. Walt Shatto relates that several breeders who were anticipating importations were invited to attend. They met at the Trade Winds Motel, Calgary. Twelve or fifteen people were present including Alberta cattle people Neil McKinnon, Laura and Byron Palmer, Bryce Stringham, Lethbridge; Mickey Collins, Winterburn; Bill and Stella Hart, and Walt Shatto, Calgary. The main business after deciding to form

The 1972 board of directors of the Canadian Limousin Association are pictured. (Front row) John Moore, president NALF, Keystone, Pennsylvania; Walt Shatto, president, Calgary, Alberta; Byron Palmer, vice president, Midnapore, Alberta; Joe Hochhausen, past president, Edmonton, Alberta; Bud McBride, Benalto, Alberta; Sherm Ewing, Claresholm, Alberta. (Standing left to right) Adrien de Moustier, Denver, Colorado; Romeo Laberge; Leoville, Saskatchewan; Stu Erwin, Winnipeg, Manitoba; Ted Godwin, treasurer, Midnapore, Alberta; Mickey Collins, Winterburn, Alberta; Jim McLellan, Langley, British Columbia. Bill Perry, Caledon, Ontario is not pictured.

an association was to register with the Canadian government, in accordance with the Canadian Livestock Pedigree Act, as a breed association. When the request was approved the group would be granted the sole right to act on behalf of the Limousin breed for the whole of Canada and function as the official registry of the breed.

Shatto, Ewing, Stella Hart, Byron Palmer, McKinnon and Collins were elected to serve as an interim Board of Directors. Absent from the meeting was Joe Hochhausen, Strome, Alberta who was also put on the board.

The previous May the North American Limousin Foundation (NALF) held its pre-organizational meeting in Denver. The NALF group was excitedly preparing for the release from the Grosse Ile quarantine station of the first Limousin bull in North America—Prince Pompadour.

He was calved as Castor by Domaine de Pompadour in France. Born January 20, 1967, he was by Baron (X Laureat and out of Baronne). His name was changed to Prince Pompadour and he was assigned registration number NIM 1. He was imported by Adrien de Moustier, Bov Import, Rimouski, Quebec. During this time in Limousin history, a group of NALF Founder members was formed. This body of 100 breeders contracted 90 percent of Prince Pompadour's first year's semen production for their own use. Funds for the memberships served as a nest egg to fund the founding administration of the NALF.

Seven of these NALF Founder members were from Canada. One of them, Sherm Ewing, Clares-

holm, Alberta, was elected as the first NALF vice president. Other Founder members from Canada included Neil McKinnon, Walt Shatto and Sam Hector, Calgary, Alberta; Nordic Farms, Christina and Lorimer Massie, Campbellville, Ontario; Dundas Farms, Cardigan, Prince Edward Island; and Mickey Collins.

In 1969 eleven bulls and five females arrived from France and became known in the trade as the "D" bulls. Six of these bulls were imported by the Brandon, Manitoba Research Station. They were Dan Pom CIM 31 (X Baron), Dudule CIM 26 (X Beau Gosse), Domino CIM 6 (X Acteur), Diese CIM 4 (X Voltigeur), Dimanche CIM 7 (X Aboyer), and Dandin C CIM 5 (X Noel). These bulls were to be used in a comprehensive research program at the station and the semen was not available for sale.

Bov Import brought in Diplomate NIM 2 (X Alaska) and Dandy NIM 3 (X Noel) at this same time. Bill and Stella Hart, Prairie Breeders, put Prairie Pride CIM 3 (X Voltigeur) and Prairie Danseur CIM 2 (X Alaska) on line in their stud service. Dr. Walter Phillips, Calgary, Alberta leased his import, Decor NIM 4 (X Beta) to American Breeders Service. These five bulls carried a heavy load of responsibility for the future of the breed. Many will say the "D" bull importation was the greatest to date.

The grand inaugural meeting of the Canadan Limousin Association (CLA) was held in March 1970 at the Palliser Hotel in Calgary. The crowd of 350

The Brandon bulls were imported in 1970 by the Canadian Department of Agriculture Research Station at Brandon, Manitoba for research and testing. At the conclusion of the comprehensive test the six bulls were sold in the 1972 Legacy Sale for a total of $307,000.

registered guests (25 of them from the United States) spent two action packed days as experts acquainted them with the Limousin breed. Research specialists outlined the breed performance tests underway. Management specialists made projections of how Limousin would fit into the scheme of future beef production in Canada.

The Canadian Limousin Association was founded at this meeting with the adoption of the Articles of Incorporation and the election of an eleven member board. Joe Hochhausen was elected president, Walt Shatto, vice president, and Laura Palmer, secretary-treasurer. Other members of the first board included Neil McKinnon, Sherm Ewing, Byron Palmer, Don Matthews, Calgary; Bryce Stringham and Leon Fouillard, St. Lazare, Manitoba; Bud McBride, Benalto, Alberta; and Paul Begin, Rimouski, Quebec.

Demand for import permits was steaming in Canada and Dr. K.F. Wells, the Veterinary Director-General for the Canadian Department of Agriculture was on hand at this first meeting to discuss the application process and answer questions regarding import procedures from the large crowd. Louis de Neuville, Limoges, France addressed the overall topic of "what were Limousin cattle like—what would they do?"

Keith McKinnon, Kenwynn Farms, Carseland, Alberta relates with this first meeting as his introduction to the Limousin breed. "I remember seeing a bull and heifer in the parking lot of the Palliser Hotel. I was extremely impressed with the muscle development; it was the most muscle development I had ever seen.

"I am a dyed-in-the-wool commercial man, looking always at the end product. When I first saw the Limousin bull, I had already been asked to breed cows on contract to Limousin. So I went to the first meeting of what was to become the Canadian Limousin Association. I ordered semen from Diplomate, Diese and Decor and used it. Charolais cattle compared to Limousin cattle were gaining more, but the

Keith McKinnon, Carseland, Alberta owns Kenwynn Farms, a sizeable commercial and registered Limousin operation. A past director of the CLA, he is presently a director of the Calgary Exhibition and Stampede. He and the McKinnon family have been Limousin supporters since the beginning—Harvey Tedford, CLA secretary-manager is pictured left.

Limousin were finishing at the right weight for the packer (1,200 pounds) in fewer days and the extra cutability was making up for the lower gain than the Charolais.''

The imports in the Palliser Hotel display Keith refers to were yearlings just released from quarantine. As one observer remembers, ''They were yearlings—definitely showing the stress of the long quarantine in France and Canada. Today we would not put them on display because of their condition. But they attracted much favorable attention at the convention.''

With the Canadian Limousin Association organization in place, there now was work to do. There were cattle to register, there was promotional work to be done and there were buyers to find.

The board of directors looked to Laura Palmer in her newly elected status as secretary to continue her efforts in coordinating these programs.

Her father, Bob Brinkerhoff, furnished an office, typewriter, phone and supplies. Laura donated her time and the Canadian Limousin Association office opened for business.

A Texas native, Laura attended Colorado State University, Fort Collins, Colorado and graduated with a degree in Agricultural Journalism. Also attending CSU was Byron Palmer, a son of one of Canada's most progressive livestock families, the Morris Palmers of Stavely, Alberta. Byron and Laura were married following school and joined the Palmer Charolais operation near Stavely.

In the spring of 1968, while in France selecting a Charolais sire for the Palmer herd, they heard about Limousin—the big red cattle of the Haute Vienne region of France. On their return home, they made further inquiries about the breed.

One of the first moves Laura made as secretary of the CLA was to publish a Canadian Association Newsletter. The first issue was published in April 1970. The mimeographed, four page, well edited first edition was mailed to 600 cattlemen reporting to them the successful Grand Inaugural meeting in Calgary. She continued to edit the newsletter until her retirement from the office in June 1972.

Her enthusiasm, cattle knowledge and leadership show in her writings, as they did in her ready response to the myriad questions which she fielded daily on the phone or in person during those formative years. She used the newsletter to inform members

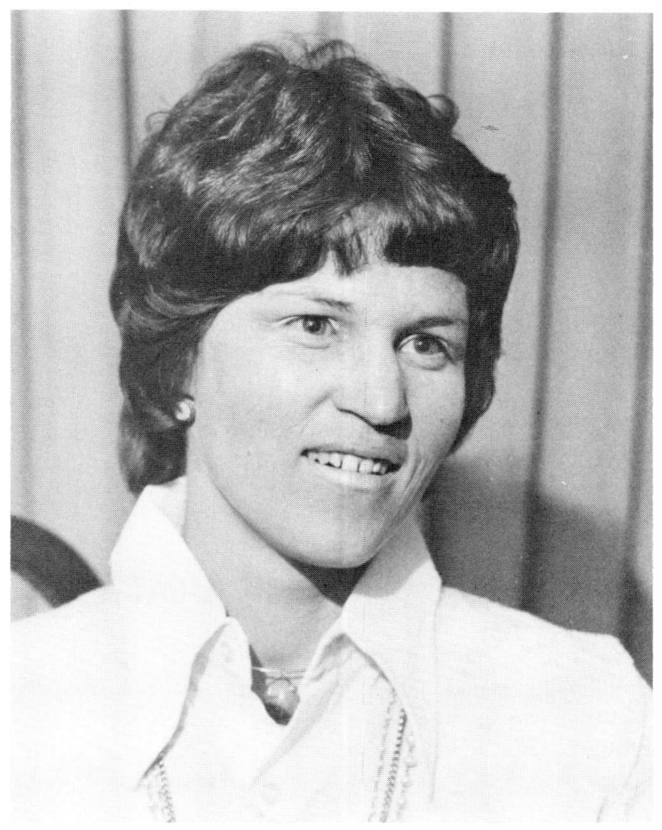

Laura Palmer, Midnapore, Alberta served as Secretary-Manager of the CLA from 1969 to June 1972. Her enthusiasm, cattle knowledge and leadership were recognized by the membership when they bestowed upon her the first honorary lifetime membership in the Canadian Limousin Association, at the 1973 convention.

on tattooing, how to fill out registration applications, what bulls were available and other useful items. She used it to chide members to get active... ''help sell this breed of cattle...sell your neighbor a membership.''

(In volume number three she editorialized...''an association (or any body for that matter) is only as strong as its members, and only exists by virtue of their interest in and support of the organization. To put it another way—put your money AND your mouth where the action is, or pretty soon there won't be any action. Nothing slows down a clear, fast-running stream like a litter of debris cluttering up the surface of the water, and nothing holds a potentially great breed back like a sluggish, do-little association backing it.''

The race to get numbers prompted many operators who had Limousin semen available to go out and

Ted Godwin (left), Millarville, Alberta served as Canadian Limousin Association president for two terms starting in 1973. Norm McNally (right), Calgary businessman and early partner with his brother Dave in LV Ranches, served as treasurer of Canadian Limousin Association. His wife Joyce was president of Limoselles for three years and was honored by the CLA with a life membership.

buy heifers, breed them and offer them for sale. The problems generated with such actions prompted Laura in the August publication to warn the normally cautious Canadian cattleman to be wary of… "hundreds of open heifers, ranging in quality from junk through passable to pretty fair, that have been purchased, pooled, and pampered just long enough to impregnate them (maybe) with Limousin semen, put a little flesh on their ribs, and offer them for sale, usually at quite a mark-up."

The warning piece was prompted by the many problems buyers encountered after they purchased "bred" females from these wholesale, put-together operations—poor breeding records, calving problems with the young, undersized and junk heifers—were just a start of the basis for complaints. The dissatisfied "new breeders" were coming back to the CLA office for assistance and to lodge their complaints about the breed to Laura.

In May 1972 Laura Palmer resigned her position as secretary-treasurer of the CLA to resume a domestic life with her husband Byron and their family on their Palmerra Beef Cattle Centre, 20 miles south of Calgary. As her successor so aptly put it in the June 5 CLA Newsletter"…without her drive and enthusiasm, it's a certain fact the CLA would not be

the rapidly expanding breed organization that we have become in the past year."

Laura has since served a term on the CLA board and was its treasurer from 1976 through 1979. She has been active on the promotion, budgeting and other CLA committees since leaving the secretary-managers post.

At the 1973 CLA convention the membership bestowed upon her the first honorary membership in the Canadian Limousin Association.

Limousin were starting to win in the 1971 summer county steer shows. The CLA committees were working hard to have a Limousin display at each of them. Live cattle, printed material to hand out and a staff for the booths had to be supplied. Promotion plans were formulated for the big winter shows coming in Edmonton, Regina, Saskatoon, and Calgary.

Numbers of Limousin cattle and members were growing rapidly. The need to gear up for processing registration and transfers by the CLA to keep up with this momentum was imminent.

In this early stage of development the CLA was using forms and procedures developed by the NALF, who were also printing the registrations and transfer papers after the CLA processed the applications. Through no real fault of anyone, problems were showing up and signs of the two associations drifting apart began to appear.

The most vexing problem centered around the NALF computers that would not compute. It created unprecedented delays in both the NALF and CLA bodies. There were halfblood heifers with three-quarter blood calves in both countries without completed registration papers.

In May of 1973 the CLA contracted with a Calgary firm to program, print and process their work. The NALF had produced over 5000 papers for them before the change.

All things were going exceedingly well for the Canadian Limousin Association in 1972. The first big event of the year was an attendance of over 200 at the convention in Calgary in February.

The CLA womens auxiliary was organized with Gloria Holt, Calgary, as its first president. The newly formed organization also selected its first Limousin queen, Jackie Chalmers, Millarville, Alberta. After a few years the national queen program was abandoned and provincial queens were elected and

One of the popular early steer champions over all breeds was the 1973 Agribition Grand Champion by Echo out of a Shorthorn cow shown by Gavin Hamilton, Innisfail, Alberta. Dr. Walter Phillips, Calgary, bought the champion for $1.70 a pound.

served in their respective areas during their reign.

Memberships were increasing rapidly. In June of 1970 the association had 126 members. It had grown to 597 by May 1973.

Canadian Provincial associations were active in Manitoba (the first one to be chartered), Ontario, Alberta, Saskatchewan and British Columbia.

The success of the 1972 convention, however, was all but eclipsed by the CLA Legacy Sale held in conjunction with the convention.

Byron Palmer was sale committee chairman and throughout the whole affair a "Laura touch" was evident. The purple coats worn by the handlers of cattle in the ring was the ultimate giveaway—the white leather halters on the cattle; the decorated backdrop; the fresh flowers too. The class affair was held in the Kinsman Center on the Stampede Grounds in Calgary.

A tremendous crowd gathered for a parade of sale cattle in the morning. The parade announcements were interrupted intermittently with applause from the crowd in support for the animals in the parade that morning. Sonny Booth, the auctioneer, describes the event as one of the most exciting sale openings he ever attended. The program was orchestrated by sale manager Jim Baldridge.

The Brandon "D" bulls were the lead items in the sale order. Dandin C brought $176,000, a North American record for a beef bull. He was purchased by International Beef Breeders, Denver, Colorado. The six "D" bulls averaged $51,000; three fullblood bull calves $16,500; four fullblood heifer calves $19,600 and two three-quarter blood heifer calves made a $4,050 average.

It was well known by all that registration and transfer papers were months in coming to the breeder, but buyers of the bulls at the sale had a big surprise. When they stepped up to the clerk's desk to settle for their cattle, they were handed a registration certificate of their purchase along with their bill of sale. The office had set up a typewriter at the sale site and completed the transfer right on the spot. It was a sensational conclusion to a most successful event.

Chapter 3

Growth of the Canadian Herd

A news release from the Canadian Department of Agriculture in October 1970 outlined the growth of "...artificial cattle breeding in Canada. About 1,100,000 or 20% of the country's cow population

Importations of European Cattle to Canada

Year of Release	No. Cattle	Charolais		Simmental		Pie Rouge		Limousin		Maine Anjou		Others*	
		M	F	M	F	M	F	M	F	M	F	M	F
1966	105	28	77										
1967	213	38	174			1							
1968	239	38	179	4	4	4	8	1					1
1969	238	18	193	3	5			11	5	2	1		
1970	664	22	464	29	38	12	16	27	33	11	9	3	
1971	618	14	218	39	160	13	34	28	59	11	9	27	6
1972	865	7	138	14	246	11	72	16	162	11	25	74	89
1973	942	10	79	16	260	2	74	1	149	3	67	76	205
1974	906	5	61	10	244	2	80	7	136	11	76	54	217
Sub-Totals	4790	180	1583	115	957	45	284	91	544	49	187	235	517
Totals	4790	1763		1072		329		635		236		752	

Other breeds include: Brown Swiss, Blonde d'Aquataine, Fleckvieh, Tarantaise, Salers, Gelbvieh, Pinzgauer, Normandie, Marchigiana and Romagnola.

were bred in this manner last spring. The proportion is an increase of about 1% from the previous year, and about 7% from 1959. The figures cover only A.I. breeding done with semen from bull studs owned by the eleven semen producing organizations operating in Canada last year, or with imported semen."

Exports of semen showed a spectacular increase, the report stated. "The upsurge was a result of an increasing demand in the U.S. for semen from the imported "exotic" breeds—Charolais, Limousin, Simmental and Maine-Anjou—and an increased demand both in the U.S. and elsewhere for semen of other breeds, particularly the Holstein.

"Overall exports amounted to 285,236 vials of semen, or more than seven times the 36,077 exported in 1968. In dollars, last year's shipments had a total price tag of approximately $3 million compared with $400,000 for 1968.

"Of the more than 20 countries importing semen from Canada, the U.S. was the biggest customer and accounted for 238,142 vials. Charolais (84,021 vials), Simmental (82,015 vials), Limousin (27,371 vials), Maine-Anjou (16,512 vials) and Holstein (9,063 vials) were the dominant breeds in semen exports to the U.S.

Total exports of semen to the world in number of vials for the principal breeds last year (1969) were: Charolais, 86,053; Simmental, 82,015; Holstein, 47,738; Limousin, 27,371; Maine-Anjou, 16,512; Angus, 6,913; Shorthorn, 5,208; Jersey, 4,702; Hereford, 3,130; Red Holstein, 2,507; Guernsey, 1,025; and Ayrshire, 760."

The report clearly indicated the wisdom of the Canadian Department of Agriculture when they established quarantine stations to process the European breeds coming to the North American continent.

Demand for mainland European beef breed imports was so great in the early 1970's in Canada that there was talk of increasing the size of the Brest quarantine station in France. The capacity of Brest was 900 head. By the end of 1974, 4,790 head had been brought into Canada. Charolais led the list of imports with 1,763. Simmental made up 1,072 head of the total and Limousin 635. In 1971 there were 618 head released from the St. Pierre and Grosse Ile stations in Canada. The numbers escalated to 865 in 1972, 942 in 1973 and 906 in 1974. During these years, 28 Limousin bulls and 506 females were admitted.

The fact that two out of every three applications for full French imports were denied for lack of quarantine space, possible lack of purpose, availability of quality import prospects or other reasons gives one an idea of just what an immense demand there was for full French imports during this period.

The great influx of imports spawned numerous A.I. studs to start in business or expand their facilities in Canada in order to handle the demand to house, collect and distribute semen. Many of them were in the Calgary area; Prairie Breeders, Calgary; Universal Breeders, Cardston, Alberta; Western Breeders, Balzac, Alberta (also White City, Saskatchewan and Woodstock, Ontario); and Southern Sires at Lethbridge. American Breeders Service also opened a Canadian unit at Bragg Creek, Alberta. BCAI was located at Milner, British Columbia and Calgary, Alberta. The large studs in Ontario expanded—United Breeders, Guelph; Eastern Breeders, Kempville; and Western Breeders were some of the

Goldenview Farms, High River, Alberta is owned by Jim and Ruth McBride. Their leadership in the show ring and breed association work has spanned 18 years. Jim and Ruth were honored with the Limousin Leader of the Year award in 1975, the first year the award was presented. He served on the CLA board and was Chairman of the Canadian Limousin World Congress. In 1977, Goldenview sold Goldenview Hugo for export to Dr. Ezequiel Tagle, Buenos Aires, Argentina. He was Grand Champion Limousin bull at the Great Palermo show in 1978.

herds. Prospects of getting more of them very soon were considered slim to impossible. Given these circumstances, many breeders wanting to get into the business looked to percentage Limousin in the United States and Canada to start or expand their herd numbers of Limousin. Those breeders setting a market with their purchases of quality and top selling percentage cattle attracted the most press attention.

Jim and Ruth McBride, who were doing business at that time in Winnipeg, Manitoba, were some of the most active buyers. They moved to Calgary in 1974 and established Goldenview Farms, High River, Alberta. Their leadership has continued to the present time.

The McBrides bought 25 head of halfbloods at the Price-McCoy Sale in Reading, Kansas. The sale report lists them as "volume buyers" in the $1,000

more active firms. A.I. Center of Quebec, St. Hyancithe, Quebec is one of the country's oldest stations. Bov Import distributed from Rimouski, Quebec.

The third breeding season was in progress for Prince Pompadour, the second for Decor, Prairie Pride, Prairie Danseur and the "D" bulls at the Brandon center. In addition, fourteen Limousin bulls were released from quarantine during this summer of 1970. Prairie Breeders announced that semen from the Brandon bulls was now available. The Canadian Department of Agriculture had issued permits for 82 Limousin for purchase in France that fall. The numbers of Limousin were growing fast but the demand was even greater.

American Breeders Service announced they had sold 44,000 doses of Decor semen from the time he was put on line in 1969 through the 1970 breeding season. Prairie Breeders had sold 50,000 ampules on their two Limousin bulls.

For the most part, few herds in Canada had more than one, two or three full French females in their

Christina and Lorimer Massie established a world-wide reputation for their Nordic Farms herd near Campbellville, Ontario. Two cows, Debonnaire and Debutante, are most respected but the list extends well beyond these two. Christina was presented the Limousin Leader of the Year award in 1975.

Jeanne Lock (pictured left) and husband Ray have JRL Limousin at Stettler, Alberta. Jeanne served as secretary of the Alberta Association for many years and was a force behind the strength of that organization. Jeanne was active in establishing the strong National Junior Shows that were held for many years at the Stettler Fair. Mel Gosling (center), Dalemead, Alberta was CLA president in 1982 and 1983. He served as the National Test Station committee chairman since 1983. An accomplished cattleman and judge, he has been called on to select many of the bulls in France for export to Canada. He has represented the CLA on many national boards and councils. Jim Lore (pictured right), Carstairs, Alberta was elected to the CLA board in 1975 and was made president in 1977. His wife June has actively served the Limoselle organization and was president for two years.

to $1,500 range. They took eight head out of the XL Ranch Sale at Bassano, Alberta. McBrides purchased the top halfblood heifer at the 1971 Dudley-Thompson Sale, Hampton, Iowa for $3,000. At

Flying Red Wheel Ranch owner, Mickey Collins, Winterburn, Alberta was Founder member #1002. Besides serving two terms on the CLA board, Mickey was a member of the 1969 interim board of directors. Mickey calved the first fullblood Limousin by embryo transplant in the summer of 1972.

Kamloops they again took the top lot at $2,400.

In 1975 they imported 17 full French females from France and started an orderly transition from purebred to emphasis on full French bloodlines. With the importation of 30 females in 1979 they became the largest herd of full French Limousin in Canada.

At the October 1970 XL Agro Beef Sale, Norm McNally, Balzac, Alberta, paid $2,550 for a halfblood bred for a three-quarter blood Limousin calf. The eight halfblood yearlings averaged $1,509 and the six calves in the sale made $844. Byron Palmer, Calgary, topped the F_1 Female Sale in Omaha, Nebraska with his purchase of a pen of seven Limousin halfbloods.

Stu Erwin, Salisbury Stock Farm, Winnipeg, Manitoba bought several halfbloods at the Dudley-Thompson Sale, Hampton, Iowa. His top was $2,200. Merle Derochie, Calgary and Sam Hector, also of Calgary, were among the list of active buyers who were setting a market for quality percentage Limousin in these formative years of the breed. With the formation of a commercial embryo transplant facility in the late 70's, they soon became Canada's largest full French herd.

With the exception of the 1972 CLA Sale featuring the Brandon bulls, all other Limousin sales in Canada paled into insignificance when compared to the 1973 CLA Legacy Sale held in February in con-

Recipients of the Annual
Limousin Leader Honor

1975: Jim & Ruth McBride, High River, Alta.
1975: Christina Massie, Campbellville, Ont.
1976: Alan & 'Sammy' Parke, Cache Creek, B.C.
1977: Dave & Sherrie McNally, Stettler, Alta.
1978: Ernie & Wilma Tedford, Estevan, Sask.
1979: Dale & Carole Barclay, Erskine, Alta.
1980: Glen & Agnes Powell, Grandview, Man.
1981: Ron & Marg Sangster, Kenton, Man.
1982: Alton & Shirley MacKay, Listowel, Ont.
1983: Fritz & Shirley Maurer, Windermere, B.C.
1984: Don & Jean Matthews, Calgary, Alta.
1985: John & Marlene Dreher, Calgary, Alta.
1986: Stan & Pat Cochrane, Alexander, Man.

junction with the annual convention in Calgary.

Some of the figures include: 60 lots averaged $5,133...1,000 folks attended the sale...five fullblood females averaged $26,900...three fullblood bulls brought a $16,833 average. A three-year-old fullblood female (Farandole) with a heifer calf at side sold for $47,000 to Dr. A.W. Stinton, Calgary.

The figures from this sale that captured the imagination of Canadian cattlemen, however, came from the percentage division of the sale. A bred 75% Limousin female sold for $8,800—three of them averaged $7,567. The 14 three-quarter blood open heifers brought a $3,618 average and 35 bred 50% females made $1,420. ''The word of this great sale spread over the country pretty fast,'' cattleman Mel Gosling, Dalemead, Alberta remembers. The breed was moving at fever pitch.

In May 1973 the government issued a change in the rules governing the exportation of European imported cattle to the United States.

1. Male imported cattle—all that have been in Canada for three years following their release from quarantine.
2. Female imported cattle—(a) Charolais females—all that have been in Canada three years following their release from quarantine. (b) Females of other breeds—all that have been in Canada for five years following their release from quarantine.
3. Males of all breeds born in Canada of two imported parents—three years following the first release from quarantine of cattle of the respective breed.
4. Females of all breeds born in Canada of two imported parents—five years following the first release from quarantine of cattle of the respective breed.

The new regulations became effective July 1, 1973.

The Limousin industry was not long in reacting. In the same CLA Newsletter announcing the rule changes, two sales featuring fullbloods were also announced. The Full French Connection Sale in Denver would sell three bulls and four females in July. The Golden Opportunity Sale, Guelph, Ontario featured four ''exportable'' full French bulls and eight females. Both sales were highly successful.

The Manitoba, Alberta and Saskatchewan associations all held fall consignment sales in 1974 and each of them had cattle bringing over $20,000.

The first surge of exports to the United States seemed to slow during the middle 1970's, only to

Harald Gunderson, founder of the Limousin Leader. *During the formative years his staff and magazine were staunch supporters and welcomed assets to the breed.*

Alton and Shirley MacKay (pictured left) were Limousin Leaders of the Year in 1982. Cedar Patch Acres, Listowel, Ontario is a family operation that includes Leslie (pictured center) and son Alan. The MacKay operation consigned the top selling bull in the 1980 CLA Bull Test. Ron Sangster, Kenton, Manitoba (center) was president of the Manitoba Association and served on the CLA Board. Wife Marg was president of the Limoselles in 1985. They were honored for their service with the appointment as Limousin Leaders of the Year in 1981. Fritz Maurer, Windermere, British Columbia (pictured right) was elected to the CLA Board in 1981. Fritz and his wife Shirley (Limoselle president 1986) were Limousin Leaders of the Year in 1983. The Maurers started shipping their Limousin calves to Alberta special Limousin feeder sales several years ago.

increase again in the late 1970's. The situation prompted concern by CLA President Jack Ward who confronted the issue in his year-end report to the CLA members. "One disturbing note," he said, "...is that we do not have enough Limousin cattle in Canada. We have imported 859 full French cattle in the last ten years, yet only 797 fullblood cattle have been registered in 1978."

In this period Nordic Farms, Ltd., Campbellville, Ontario, owned by R.H.L. Massie and Christina Baumann Massie, was the largest pure French herd in North America. The strength of its quality was revered by breeders in Canada and the United States. This firm strongly supported the Ontario association and its activities in performance bull testing, the Golden Opportunity association sales and the Royal Winter Fair in Toronto. Christina Baumann Massie is considered by many as the breed's most ardent promoter. Her efforts extended beyond Canada and the United States into Mexico, Cuba and most of the countries of Central and South America.

By May of 1973 the CLA sported 1,181 memberships and had registered 6,900 Limousin. Canada's first fullblood out of imported stock was born at Bov Import, Rimouski (bull calf Fantastic by Dandy out of Danseuse). Mickey Collins calved the first fullblood Limousin by embryo transplant at his Flying Red Wheel Ranch in the summer of 1972.

Breeders attending the 1974 CLA Convention in Calgary awoke to find a printed report of the previous day's Limousin activities under their hotel room doors.

It was the introduction of the *Limousin Leader*—Canada's Limousin breed publication—that was to enhance the breed promotion in the coming years. The four page tabloid under the door of each CLA registrant was the first of many spectacular moves which Harald Gunderson, owner-editor of the monthly, made in the next ten years that he owned and published the *Leader*.

Harald's interest in a Limousin publication came after he became an investor in Limousin as a member of Acme Limousin Breeders, Ltd., Calgary. One of the firm's purchases was the top selling $41,000 lot at the Calgary Sale the day before this new publication appeared. The *Leader* replaced the CLA Newsletter and for the first months was subsidized by the association. Its popularity grew rapidly and by March 1975 the paper had expanded to 28 pages.

An experienced editor, Gunderson outlined the publication's direction in the April 1974 editorial. He stated, "...will bring you the news of the Limousin breed each month. We think this news will be topical and current, designed to keep the Limousin breeder informed on things that are happening with Limousin."

Readers of the *Limousin Leader* quickly learned about Harald's love of humor in his first editorial outlining the objectives of the publication. "It is hoped that we won't take our cattle problems and ourselves too seriously...we will try and keep in mind that people will pay more for the privilege of laughing than for practically anything else."

As an astute editor and manager he seemed to enjoy controversy. Perhaps this aspect of Harald as an editor was best explained by him when he conducted a communications seminar for the association in November of 1979. He advised communicators attending the seminar, "...The editor must set forth his views with as much force and clarity as he can muster and in a manner calculated to have the greatest possible impact on the mind of the reader.

"His task is not to please, but to prod; not to insulate his views, but to stir people up and make them think for themselves.

"He must afflict the comfortable, even as he comforts the afflicted. The editor who begins to worry about his popularity might as well turn in his typewriter."

The enthusiasm, dedication and activity of Harald and his staff at practically every Limousin event added materially to the growth and success of the Limousin breed in Canada.

One of the popular activities sponsored by the publisher is the annual Limousin Leader of the Year award presented during each CLA convention. The first award was presented in 1975 to Jim and Ruth McBride and Christina Baumann Massie for their "outstanding contributions to the Limousin breed in Canada through enthusiastic promotion, high integrity, sincere dedication and leadership in Canadian Limousin circles."

Harald sold the *Limousin Leader* in June 1984 to Randy Bollum, who served as the publication's editor since 1980. Harald continues to publish *World of Beef and Feedlot Management,* a national all breeds publication and still breeds Limousin cattle in cooperation with his son-in-law, Charles J. Watson of Mount Forest, Ontario.

Chapter 4

Promotion of Canadian Limousin

A great promotion point for the breed has been Limousin's complete dominance of carcass and live steer shows. It is rare that a Limousin crossbred steer does not win a carcass show. They are heavier muscled steers with adequate fat cover and yield a carcass that is the envy of the beef business. This has carried over into the judging of the live steers as well.

Harvey Tedford, secretary-manager of CLA relates to the importance of these shows. "The primary trait that has made Limousin probably the fastest growing breed in Canada is this carcass superiority. Limousin crossbred cattle ideally fit the Canadian beef grading standards with maturity and backfat thickness per 100 pounds of carcass weight which are the major factors governing whether a carcass can be graded A-1, Canada's top grade. Unlike USDA standards, marbling has little bearing on the grade; this allows for generally leaner, higher cutability beef to receive the top grade—hence the top price."

The early wins in the formative years of the breed were especially effective because they were such sensational wins. Many people had not seen a Limousin crossbred and here was a grand champion. One thing was certain, as soon as Limousin steers started going to the shows in 1971, they started winning.

The two biggest Junior shows that summer in Canada were in Calgary and Edmonton. The wins were spectacular. There were six Limousin crossbreds among the 429 steers in the two shows and Limousin came away with both grands and one reserve.

There were only two halfblood Limousin steers in the 350 head 1971 Calgary and District 4-H steer show. Kirk Gosling, Dalemead, Alberta showed one to the grand championship position over all breeds, and Marilyn Huggard, Dalemead, showed the other to the reserve grand position. Kirk's father, Mel, selected both calves out of Neil McKinnon's 1970 calf crop at his Bassano, Alberta ranch. The spectacular win at Canada's largest 4-H show provided a major boost for the breed.

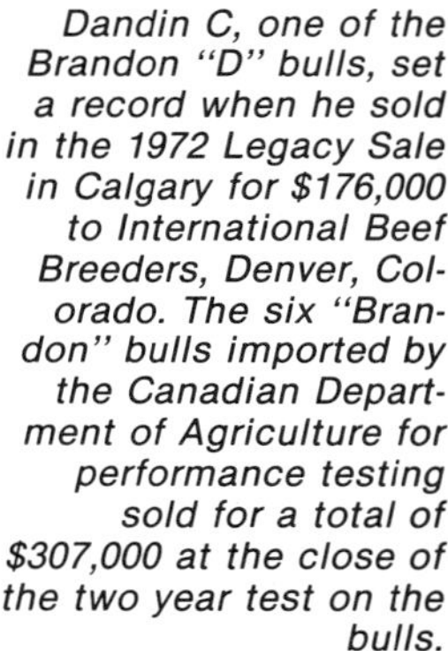

Dandin C, one of the Brandon "D" bulls, set a record when he sold in the 1972 Legacy Sale in Calgary for $176,000 to International Beef Breeders, Denver, Colorado. The six "Brandon" bulls imported by the Canadian Department of Agriculture for performance testing sold for a total of $307,000 at the close of the two year test on the bulls.

The other heralded win was Lynn Armistead's grand champion and supreme champion market steer over all breeds at the Western Canadian Stock Show at Edmonton, Alberta in April. The son of Prairie Pride was bred by Bill Hebson, Okotoks, Alberta.

At this Edmonton show only four Limousin cross steers were exhibited in the 170 head show. In an unprecedented situation all four of them were in the ring for the championship show.

John R. Lockhart was appointed Secretary-Manager by the CLA board in June 1972. Born and raised on a livestock farm in southwestern Alberta, he attended high school at High River. He was an active 4-H member before going to the University of Alberta where he graduated with a B.S. in Agriculture and a major in Animal Husbandry. Prior to his appointment as manager of the CLA he was acting head of the commerce department in a 1,200 pupil high school at Kitimat, B.C. He had extensive experience in agriculture and agri-business, farming, 4-H extension work, teaching and working as a radio commentator and agro-meteorologist. He and his wife Mae have two sons, Ken and Douglas, born in 1965 and 1966 respectively. Both have been most successful with their 4-H projects.

The CLA manager's report to the membership in August listed more Limousin successes…"late May, June and July saw Limousin crossbred steers win with consistency in 4-H shows across the prairies but primarily in Alberta: calves shown by Darrel Hay, Wetaskiwin, reserve grand; Lorne Kelndorfer, Killam, grand champion; Fiona LeGuard and Steven

Calihan, Foothills (High River), grand and reserve grand champions; Glen Summers, Calgary, grand; Wayne Thomas, Taber, reserve grand; Ron Hoar, Red Deer, reserve grand; Mark Secord, Edmonton,

John Lockhart was Secretary-Manager of the Canadian Limousin Association from 1972 into 1975. His keen foresight and diligent work laid most of the base for the successful computer programs that were later finished and serve the association well today.

reserve grand; Kristin Bohrson, Saskatoon, reserve grand; and Peter Bertolotti, Kamloops, reserve grand. Not to be outdone, Linda Boire, Saskatchewan Limousin Princess, had her heifer declared grand champion of the Estevan District Fair. A Limousin steer was grand champion at the Ottawa Winter Fair.

"During September, October and November, Limousin steers began winning the prestige shows in the country with Jim Laird of Kamloops winning grand champion carcass at the P.N.E.; Gordon Crawford of Glencoe, Ontario, reserve grand champion carcass at the Toronto Royal Winter Fair; and Ken Traynor, Delisle, Saskatchewan, grand champion carcass at the Regina Agribition. Limousin steers have also been winning group competitions; Bud McBride won the pen of five at the Regina Agri-

Alan and Sammy Parke, Cache Creek, British Columbia were Limousin Leaders of the Year in 1976. Alan served two terms as CLA President where he saw the National CLA Bull Test Station become a reality. Their Bonaparte Ranch is the oldest ranch west of the Rockies in continuous family operation.

bition with his black sons of Prairie Pride, while Gavin Hamilton won the get of sire at the Royal Winter Fair with his three red sons of Decor. And saving the best for last—Lynn Armistead of Onoway, Alberta exhibited another son of Prairie Pride bred by Bill Hebson of Okotoks to the grand championship at the Royal Winter Fair. Their 1,140 pound steer later sold for $3 per pound to Dominion Stores of Canada."

Limousin steers started winning as soon as they started showing...and they are still winning. The 1971 and 1972 show successes were only a start. The impact of winning steer market classes was one thing, but the domination of carcass contests by Limousin sired carcasses since the early days to the present time has been phenomenal and never before documented.

The meat chain in Canada was not long in finding that these Limousin carcasses fit into Canada's grading system perfectly. The large ribeye, the thin, uniform fat cover and high overall cutability gave them the edge over the other top grading steers of the industry.

Don Matthews, Calgary, Alberta was inducted into the Canadian Agricultural Hall of Fame in 1985. He served as president of the CLA and has represented the CLA on several national committees.

The Highland Stock Farms tradition as seedstock producers is being carried into the third generation by Rob Matthews. Wife Marci is pictured left.

Most fed steer and heifer carcasses graded in Canada fall into the top four grades A1, A2, A3 and A4. Most times of the year there is very little price differential between these grades. The top grade is A1 with the A2's often selling at the same price. The recent changes made by the Agriculture Canada Grading Services effective January 1, 1986 bring the B1 and B2 grades closer to the top grades A1 and A2.

The numbers in the system score the fat over the ribeye as measured at the 12th rib. The fat thickness correlated with the grade is as follows for the A grade:

Canada A1— 4 mm (.1576 in.) - 9 mm (.3546 in.)
Canada A2— 9 mm (.3546 in.) - 14 mm (.5516 in.)
Canada A3—14 mm (.5516 in.) - 19 mm (.7486 in.)
Canada A4—19 mm (.7486 in.) - over

The letters of the grade refer to the other meat quality characteristics—fat color, meat color, maturity and muscling. For practical purposes we can say that the B1 and B2 graded carcasses have the same color of fat and lean, youthfulness with good cutability as compared to the A1's and A2's.

Fat cover over the rib for a B1 grade is 2 to 4 mm (.0788 to .1576 in.). Generally the underfinished carcass is relegated to the B grade. The B1 and B2 specifications were raised in the recent grade change. The grade expects these grades to approach the A1 and A2 prices even closer in the future.

It is uncommon for a carcass championship to go to a breed other than Limousin in any of Canada's major shows. The best known of the string of championships over all breeds is the Agribition record of 11 times in a row. At the 1986 event in Regina, Saskatchewan the grand champion, the reserve grand champion and the next 15 places were all Limousin. In 1984, seven of the top ten placings were Limousin.

In 1981 Limousin made a sweep of carcass championships at the winter shows. At the Brandon Ag Ex the grand and reserve were Limousin (top seven places); Edmonton Farm Fair, grand and reserve (top five placings); Regina Exhibition, grand (five of the top six placings); Toronto Royal, grand and reserve (seven of the top ten); Calgary Round-Up, grand (six of the top eight).

For many years the CLA has paid additional premium money for Limousin cattle (live market steers and carcass steers) that were grand champion

Kevin Stanton, Seebe, Alberta (pictured left) has been a staunch worker for the Alberta Limousin Association and its activities. Bernie Payne, Lloydminster, Saskatchewan served on the CLA board and as president of the Alberta Limousin Association.

Wilbur Stewart, Big Valley, Alberta conducted his own carcass research in 1977-78. He bought a lot of half and three-quarter blood (some seven-eighths) steers at the Olds, Alberta auction and fed them to 1,262 pounds. With the cooperation of XL Beef, Ltd., of Calgary and retailer Merite of Montreal, the 108 carcasses were followed all the way through the chain. Ninety-five percent graded A1 or A2. They dressed 63%. Merite sent glowing reports on the desirability of the high cutting shipment.

and reserve grand champion at one of the 12 to 15 major shows approved in advance by the CLA Board of Directors. The premiums have been $100 for grand champion and $50 for reserve.

CLA Secretary-Manager Harvey Tedford pictured with dad Ernie, Estevan, Saskatchewan. Ernie and Wilma Tedford were honored as 1978 Limousin Leaders of the Year for their leadership in the Saskatchewan and CLA organizations.

In order to qualify the exhibitor must send a picture of the champion plus a copy of the entry form showing the breed of sire and the full name and address of the exhibitor to the CLA office. This premium program has been costing the association from $2,000 to $3,500 a year.

Some of the records made by Limousin that have been reported since 1979 include:

Grand champion carcass at Agribition for 11 consecutive years. Seven of these years they also had the reserve grand champion carcass; The grand champion market steer at the Regina Exhibition nine of the last 14 years—reserve three years; At the Calgary Round-Up Limousin won five of the last seven market steer grand championships; In the last nine years at the Royal in Toronto Limousin carcasses were grand six times—reserve six times; The last two years in a row at the Royal, Limousin had the grand champion pen of five and pen of 10 market steers— they also had the reserve champion pen of each class one of these two years; In 1978, 20 of the top 35 carcasses were Limousin at the Royal, 25 of the top 30 in 1979 and seven of the top 10 in 1981; At the Calgary Round-Up, Limousin carcasses were grand champion in 1981, 1982, 1983 and 1984—also reserves in 1982, 1983 and 1984; At the Brandon Ag Ex Limousin carcasses have won six of the last seven years, reserve three of these years also.

Harvey Tedford was appointed secretary-manager of the Canadian Limousin Association in November 1975 after serving as the organization's assistant secretary-manager for a year prior to this appoint-

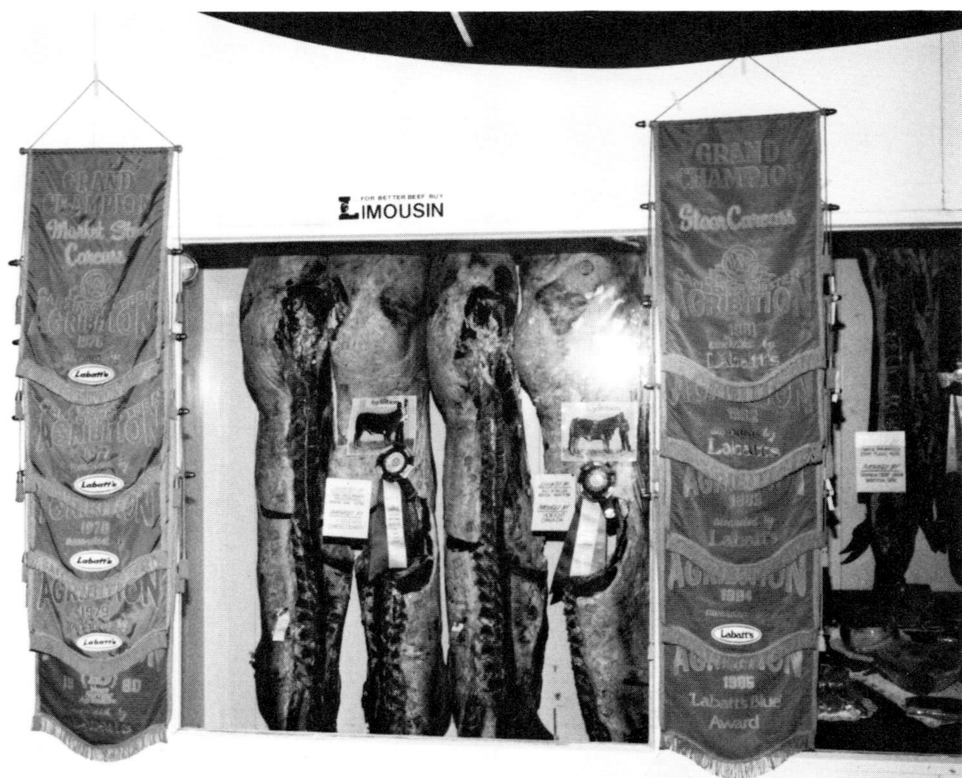

For 11 consecutive years, Limousin carcasses have won the carcass competition at the Canadian Western Agribition, Regina, Saskatchewan show. At the 1986 show the next 16 places below the grand champion were also Limousin sired carcasses. In 1985 the top 12 places were Limousin in the 54 carcass event.

ment.

Harvey is committed to the breed that "has advantages for all segments of the cattle industry."

Born and raised on a livestock farm near Outram, Saskatchewan, Harvey attended school in that community 18 miles west of Estevan. He graduated from Estevan High School and attended the University of Saskatchewan. He later transferred into the faculty of agriculture on the Saskatchewan campus where he received a bachelor of science in agriculture and a major in animal science. His thesis was an "Evaluation of Breeds and Crossbreeds."

In March of 1972 Harvey became the feed formulator for Federated Co-op in Saskatoon. That organization transferred him to Edmonton where he became their sales representative for northern Alberta.

Harvey is the son of Limousin breeders Ernie and Wilma Tedford, Outram. Ernie has been an active member (and president) of the Saskatchewan Association. Harvey and wife Dawn live in Calgary with their family, Catherine, James and Melody.

One sees the firm confidence Harvey has for the breed and its usefulness for each of the segments of the beef chain by his daily contacts and writings. It shows in the believability of the ad copy and video presentations he writes for association promotion.

Commercial cattle prices were in a down market when Harvey assumed his new position. In 1975 registrations and transfers leveled out, new mem-

bers continued to gain but problems with computers producing registration papers persisted. By the time Harvey issued his 1976 annual report he was able to report a nine day turnaround for work coming into the CLA office.

Growth in transactions (registrations, transfers, new members), however, was slow through 1979 and created a real cash flow problem for this period. In 1980, the growth started and by 1984 transactions had doubled. In 1986 registration of full French cattle was 3.82 times that number registered in 1979.

In spite of the disastrous cattle market of the late '70's, prices Limousin breeders received for their bulls was not as bad as other breeds experienced. Even with the distressed market, commercial Limousin steers and heifers were bringing a two to ten cent premium at the auction. This premium in many instances brought cattle operations from the loss column to break even status.

In 1982, the CLA sponsored the Canadian World Limousin Congress which was held in conjunction with the Calgary Stampede, July 9 and 10. The two day show attracted 253 entries from North American breeders. Berwyn Wise, Irricana, Alberta was granted the honor of judging the prestigious event.

Grand champion bull honors went to Goldenwest Farms, Miami, Oklahoma and grand champion female ribbon was picked off by Bodell Limousin,

Sherwood Park, Alberta.

The 59½ lot sale held in conjunction with the event averaged $8,763. Nine fullblood bulls offered in the sale averaged $7,668 and 11 fullblood pairs averaged $11,195.

A tour planned as an attraction to the World Congress started in Toronto with international guests arriving on the 5th of July. Stops on the tour included Niagara Falls, a winery, the campus of the University of Guelph and visits to Nordic Farms and other Ontario Limousin herds. In Calgary, they watched the Stampede Parade, visited Goldenview, Kenwynn, Golden Limousin Acres and other Limousin herds in the vicinity. After the World Congress festivities at the Exhibition grounds, the tour proceeded to Highland for breakfast and then on to Banff and Lake Louise. A "good ol' western style barbeque, a barn dance, chuckwagon breakfasts," viewing chuckwagon races, a gondola lift, hayride and many other special events kept the guests on a busy schedule. Guests from 10 countries joined the Congress tours and festivities.

The district, regional and national shows have been well supported by Canadian breeders. Consequently, they have been strong promotional events for the breed as well as for the breeders who participate.

Canada's largest 1986 Limousin breeding cattle show was held at the Northlands Farm Fair in Edmonton, Alberta. The mid November show is held in the 250,000 square foot Agri Com building completed in 1984. Market steers and carcass classes are a part of each year's events.

In the last week of March, Edmonton hosts the Northlands Western Canadian Stock Show during which the various breeds hold bull sales. A hog show, rodeo and a machinery show are held in conjunction with the bull exhibits and sales.

One of the largest shows in Canada is the annual Canadian Western Agribition Show held in late November of each year in Regina, Saskatchewan. With the large Agridome completed in 1985, the show has widened its scope and classes for the Limousin portion of the show. It now has the standard Limousin breeding classes, open steer and carcass classes, pens of five and 10 commercial fed steer classes as well as pens of five and 10 replacement heifer classes. They have also added pens of three commercial bull classes that compete for interbreed

Marvin Latimer (pictured right), Hayard Farms, Innisfail, Alberta, has always been a willing worker on the National Bull Test and other CLA and Alberta association committees. He is establishing a reputation with his polled Limousin. Long time Limousin breeder Lynn Combest from Erskine, Alberta is a regular at most of the major shows.

competition. Kent and Myria Holland of Dilke, Saskatchewan exhibited the grand champion pen of five replacement heifers in 1985 and 1986.

Before the 5.7 million dollar building project to put the show under more roof started in 1982, the show attracted 1,200 producers with 4,500 head of livestock. A total of $2.4 million in livestock sales to domestic and foreign buyers was held in conjunction with the 1982 event. In 1986 there were 20 purebred cattle sales and 11 commercial livestock sales held at the 700,000 square foot complex.

Canada's oldest and most highly respected show is Toronto's Royal Winter Fair held each year in mid November. Limousin breeding classes were first added to the show schedule in 1980. The support of Christina and Lorimer Massie along with the Ontario Association in getting the classification is noteworthy. The first show had 104 head lead out for judging.

By the time Limousin breeding classes were admitted to the Royal cattlemen were well aware of the breed. As early as 1974 the grand champion steer

REGISTRATIONS	1978	1979	1980	1981	1982	1983	1984	1985	1986
FULL FRENCH	819	859	1,094	1,149	1,560	2,074	2,031	2,268	3,285
DOMESTIC PUREBRED	4,013	4,034	780	1,186	2,155	2,955	3,611	3,910	3,809
PERCENTAGE	85	369	4,173	4,118	4,372	4,507	3,919	3,249	2,644
TRANSFERS	4,266	3,221	3,778	3,620	5,335	7,049	7,483	7,857	7,960
TOTAL	9,183	8,483	9,870	10,073	13,422	16,665	17,044	17,284	17,698
ACTIVE MEMBERS	—	—	110	159	160	225	232	240	200

was a Limousin. In 1975 the grand and the reserve grand champion carcasses and the pen of five champion fed steers were Limousin. In 1978 the top 20 steers in the carcass contest were Limousin, which included the grand and reserve.

Other regional fall shows where Limousin breeders vie in numbers are Ag Ex at Brandon, Manitoba and the Lloydminster show in Saskatchewan.

Leading summer shows include Stettler, Alberta where the Canadian junior heifer show had been held the end of July each year; the Western Fair held in London, Ontario in mid September, the P.N.E. held annually over Labor Day at Vancouver, British Columbia and the Calgary 4th of July Stampede. The Juniors now hold their national show at Agribition and the Royal on alternating years.

In 1984, Randy Bollum purchased the *Limousin Leader* from founder owner Harald Gunderson. President Stan Cochran, in his report to the membership, wished to "recognize Randy Bollum as the new owner and publisher of the *Limousin Leader*. Randy is an aggressive and very enthusiastic young man and we are happy to have him on the Limousin team."

Randy was born and raised on a farm near Blue Earth, Minnesota where his family breed Limousin cattle. While in 4-H and FFA in the United States he was actively exhibiting and promoting Limousin cattle in many shows and sales in the upper Midwest of the U.S.

Randy is a 1979 graduate of the University of Minnesota where he majored in Animal Science and pursued several courses in communications. He served as president of the North American Limousin Junior Association for two years. While at the University of Minnesota he won many awards through his

The 1982 Limousin World Congress in Calgary was closed with a successful auction of 59½ lots to average $8,763.

participation on the University general livestock, livestock evaluation and meats judging teams.

The monthly publication continues to be "the Canadian voice of the Limousin breed." The in-depth reporting of shows, sales and other Limousin events serves the breed and association well.

Chapter 5

The Search for Commercial Advantage

The Canadian Limousin Association has historically had a watchful eye toward what they define as "the needs of the commercial cattleman." Two CLA and provincial programs have substantially aided these goals. One has been the national bull test and the other the Limousin feeder calf sales program. In one the CLA is helping identify the breed's most useful bulls and in the other helping the commercial man to merchandise his feeder calves by Limousin bulls. Each of the programs is directed toward influencing and attracting this user of their product, Limousin bulls.

Harvey Tedford, in a published article in the *Limousin World* in 1985, pointed out that..."the initial breeders that formed the Canadian Limousin Association and the provincial associations that followed were generally composed of ranchers, farmers, and feedlot operators. Their practical, long term views were formed with the idea in mind of developing a strong foundation based on credible records and an elected group of representatives dedicated to serving its membership. Although it is difficult to generalize, and especially to find the actual figures, it appears that Canadian Limousin breeders have a tendency to stay in the registered cattle business for more years than do breeders of other purebred cattle in Canada. Most of the people involved in Limousin, and nearly all of the most influential ones, make a large part of their income from their Limousin operations. They are dedicated cattlemen with long term objectives. Their marketing and breeding strategies reflect their commitment and sincerity."

The feeder sales are annual events that start with the fall calf run in October and go strong throughout November and into December. Their purpose is to gather large numbers of Limousin feeder calves so buyers can buy and load cattle liners or rail cars for the designated feedlots. It is not unusual for buyers to travel 2,500 miles for a sale or group of sales and buy several carloads.

As early as 1972 the CLA Newsletter was alerting Alberta breeders that a Limousin feeder calf sale was being scheduled in the fall. They were asking members to contact commercial cattlemen and solicit their Limousin crossbreds.

In the beginning the CLA and or the provincial associations were quite active in promoting the feeder calf sale program. They encouraged cooperation from the local auction markets, solicited ranch-

Sherrie and Dave McNally, LV Ranches, Erskine, Alberta have been some of the hardest workers in the CLA and Alberta associations. LV bred the high performing and top selling bull in the first National Bull Test. They were 1977 Limousin Leaders of the Year. Dave operates the Canadian Cattle Consultants sale management firm.

The Powells of Grandview, Manitoba have made Glenkair Farms a famous name in North America. Patriarch Glen (center) is receiving congratulations from CLA president Alan Parke (left), owner of Bonaparte Ranch, Cache Creek, British Columbia. Son Bob, a member of the CLA board, is pictured right.

ers to consign to the special sales and contacted eastern buyers. For a time the auctions and CLA ran cooperative ads on the circuit of sales.

As time passed and the auction operators saw the advantages of a ''circuit,'' the CLA involvement lessened. They still serve as a hub for coordinating dates to avoid overlapping scheduling. The real story behind the success of the feeder cattle sales is the fact that feeders are willing to pay a premium for Limousin crossbreds. They will pay this premium if they can buy them in numbers so they can feed and market them to a full advantage.

''The hottest demand has been for uniform groups of solid red steers showing lots of 'Limousin' characteristics—usually these are three-quarter blood calves,'' Tedford says. ''This supply of three-quarter calves will increase as commercial producers add foundation halfblood heifers to their base herds and the supply of fullblood and purebred bulls increase.''

In 1986 there were over 9,300 juniors enrolled in 4-H beef projects in Canada. To accommodate this buyer's market, many feeder calf sale operators permit buyers to make one or two picks out of a draft as it sells.

Some of the largest 1986 Alberta sales, as reported in the *Limousin Leader,* included 1,200 head at Highwood, High River; the Volds at Ponoka reported, ''this year the quality and numbers were up,'' and they had 1,800 in 1985; Stettler had ''3,000 head of Limousin and Charolais calves with one pen of 13 Limousin weighing 597 pounds bringing $108/cwt''; Veteran with 1,000 head reported one lot of twenty-one heifer calves averaging 450 pounds sold at $124.25/cwt; and the Balog market at Lethbridge features a two day crossbred sale.

In Saskatchewan, Limousin breeder Kelly Yorga, Flintoft, coordinated five Limousin feeder calf sales with 3,500 head consigned in sales at Weyburn, Moose Jaw, Saskatoon, Maple Creek and Swift Current. The 300 to 450 pound steers averaged $110 to $120/cwt and heavier calves went from $105 to $110/cwt.

Gerry Smailes, Hanover, Ontario, sold 1,400 to 1,500 head in his October 27 sale last year. Prices ranged from $100 to $120/cwt with a top of $127. Smailes breeds his commercial herd of Angus and Charolais x Angus cows to Limousin bulls.

Randy Bollum is owner publisher of the Limousin Leader, *a class publication dedicated exclusively to the Limousin breed.*

Clayton Curry, Curry Cattle Company of Sunnynook, Alberta holds a unique feeder calf sale each spring. The order buyers attend the ranch to make their purchases and buy the Limousin calves before they suffer the stresses involved in transporting them by rail or truck to market destinations. Most of the 1,300 steers and heifers sell into Ontario feedlots.

One of the most successful programs to be sponsored by the CLA has been the Canadian National Limousin bull test, now in its eleventh year. Breeders from across the country annually consign their best calves each fall to the test where each is compared with over 100 other test bulls in a 140 day performance test. Upon arrival at the station, the bulls are penned according to age to begin their test.

John Knight conducts the test in his specially designed lot, just 9 miles east of Calgary.

The first 28 days of each test program are used to provide an adjustment period to allow all the bulls to recover from the stress of hauling. After the initial 28 day adjustment period, the bulls are officially weighed for beginning test weights, and are reweighed every 28 days. The growth rate in each 28 day period is then used to calculate the indexes for gain and a weight per day of age for each individual. The combination of these growth measurements is then used to calculate a rank based on their total performance.

Each year the National tests three different categories of bulls based on their percentage of Limousin blood. Domestic purebred bulls consist of all bulls of 90 percent or greater Limousin blood derived from an upgrading program using cows of a second breed. The percentage bulls comprise a second category and consist of all bulls less than 90 percent Limousin blood, in this case mainly three-

Gerry Good, Carstairs, Alberta was elected CLA president in 1986 and again in 1987. During his two terms as president of the Alberta Limousin Association, he actively promoted special Limousin feeder calf sales at the various provincial stockyards.

Carole and Dale Barclay, Erskine, Alberta are pictured with 1978 CLA President Jack Ward, Arrowwood, Alberta. Barclay served two terms as CLA president. In 1979 the Barclays were made Limousin Leaders of the Year.

entire duration of the program.

The top indexing bulls are screened by a strict selection committee and the qualifying bulls are offered at public auction every April in what has become one of the top sales of the year. It has been encouraging to see the repeatability of the test as several of the top indexing or top selling bull's sons come back to perform well in successive years.

The bulls on test are fed a medium energy grain ration, consisting of 20 percent barley, 65 percent oats, 7 percent of a 32 percent beef supplement, 5 percent beet pulp and 2 percent molasses. The ration is fed at a ratio of approximately two parts grain to one part hay, twice each day. The ration is designed to challenge the genetic differences for growth potential and express them without harming the bull's future reproductive capacity.

Over the years, visitors to the test station have had the opportunity to evaluate, not only the feeding program and the selection data, but the uniformity and the obvious carcass strengths of the Limousin breed.

The Canadian Limousin test program is a reference, not only for the visitors, but for breeders who are able to observe a cross section of sixty different breeding units from across Canada. This provides breeders with the opportunity to compare the performance level in their herd with their fellow breeders. The bull testing program was originally conceived as a merchandising tool to provide top performance tested Limousin bulls for the commercial man. The fact that more and more purebred operations each year buy herd bulls at the sale attests to the soundness of the program.

quarter and seven-eighths crosses. Purebred and percentage Limousin have been developed through intense selection at each generation and will eventually make up the bulk of Canadian herds. It is through this program that strains of polled Limousin cattle are being developed. The third category of bulls tested is comprised of fullblood bulls.

Great care is taken to check and recheck each measurement. As bulls start and finish, the test weights on two consecutive days are taken and averaged for the official on and off test weights. All information is carefully compiled and sent to each breeder and potential buyer upon completion of each segment. The breeders and buyers are then able to follow the performance of individual bulls over the

The following brief history of the establishment of the national test and the results of its first ten years has been supplied for this history by Canadian Limousin Association secretary-manager, Harvey Tedford.

The Canadian Limousin Association finalized plans to have a Canadian Limousin Association Bull Test Station in a Board Meeting in March of 1977.

Several feed lots and test stations were toured by Alan Parke, the president of the Canadian Limousin Association and the Secretary-Manager, Harvey Tedford. As a result of the tour and the subsequent presentation to the Board, the Committee was approved and expanded. The Bull Test Committee was formed under the Chairmanship of Dr. Ted Burnside to include the following members: Dale Barclay,

The national Canadian Bull Test has been one of the most successful enterprises of the Canadian Limousin Association. The 110 head lot is conveniently located 10 miles east of Calgary just off the Trans Canada Highway. After the test between 40 and 50% of the bulls are selected to sell in the sale held at the Stampede grounds. Bulls are tie broke, dehorned but not clipped.

Marvin Latimer, Wilbur Stewart, Dave McNally, Alan Parke and Jim Lore. It was through their efforts, recommendations and direction that we established the rules and regulations which are now a part of the Canadian Limousin Association Bull Test Program.

The objective was to set up a centre to benefit the breed by providing an opportunity for the breeders to display their bulls at one central location, under top management and with maximum exposure. Breeders would then be able to select from this group of bulls, individuals on the basis of performance and conformation which will be the leading herd sires in years to come.

This was believed to be the first National Bull Test Centre of any beef breed in Canada. All other bull testing programs consisted of bulls from a local area or provincial group. The Canadian Limousin Bull Test Station was a program which crossed Provincial boundaries and was open to any member of the Canadian Limousin Association.

We encouraged breeders of the day to nominate bull calves from the top end of their herd which exemplified their breeding program and would be prospective herd sires within the breed. All of the bulls were to be born between January 1st and April 30th and were sent to the station in the fall of each year.

A health program was set up to ensure that strict attention was paid to a preconditioning program to minimize stress on the bulls and assist them in the adjustment period.

The first Canadian Limousin Association Bull Test Sale was held at the Calgary Exhibition and Stampede Grounds on April 17, 1978. The first test station consisted of 75 head of Limousin bulls from 42 breeders and five provinces. The bulls were indexed by a computer system developed by Dr. Larry Schaeffer of the University of Guelph. The bulls were indexed to account for the large age spread by

Wilbur Stewart (left), Big Valley, Alberta breeder, farmer, rancher, held his 11th annual Unsurpassed Sirloin Production Sale in 1987. His Imperial Ranch is one of Alberta's largest. He is owner of the high performance sire Pub and serves as Canadian Limousin Association treasurer. Walt Shatto (right), Calgary, Alberta was Canadian Limousin Association president in 1972 and served on the NALF board. Starting as a partner in Limousin Breeders of Canada and Acme Breeders, he is still an active breeder in Molalla, Oregon.

Gerry Long, Mainline Limousin, Lambeth, Ontario has been an effective leader in Ontario Limousin activities including getting the Ontario Fieldman program on track. Pictured is wife Renie and Susan and Tim Dunlop who figured prominently in the Nordic successes before joining Mainline.

weaning weight with a repeatability of 74 percent. His yearling weight proof was a plus 21 pounds with a repeatability of 68 percent, which was the highest of all of the bulls tested in the program. He has gone on to rank in the top 10 percent of all Limousin bulls in the breed tested on the Canadian Sire Monitoring program.

The second Canadian Limousin Bull Test Station Sale was held on April 16, 1979 at the Calgary Exhibition and Stampede Grounds. A total of 44 lots were sold from a total of 111 tested. The high testing bull was Mr. Eclaireur with an average daily gain of 3.13 pounds who sold for $5,000 to Avid Enterprises of Acme, Alberta. The high selling bull was Kensington who sold for $15,200 to Keith McKinnon of Carseland, Alberta and Mel Gosling of Dalemead, Alberta. Kensington stood second in the 140 day test with an average daily gain of 3.11 pounds per day. Kensington later went on to be progeny tested and was also proven to be a plus bull for wean-

adjusting for the age at the start of test and using a regression approach to give a more precise estimate of each bull's average daily gain on test. The bulls were fed a medium growing ration at 68 percent TDN and 13 percent protein.

The first forty head of bulls sold from the first Bull Test Station program on April 17, 1978 for a gross of $96,325 and an average of $2,408.

The high selling bull sold for $10,400 to R.A. Chiswell of Lacombe, Alberta. This bull was consigned by L.V. Ranches of Stettler, Alberta. He was an Eclair son who had an average daily gain of 3.24 pounds per day and a weight per day of age of 3.09 pounds. His index for gain was 103.3 and for weight per day of age 105.0. Limac Joe was later tested in a progeny test program by the Canadian Limousin Association and proved to be plus 25 pounds for

Canadian Limousin Association secretary-manager Harvey Tedford (left) is pictured with Berwyn Wise (right), Irricana, Alberta. Berwyn managed the Canadian Limousin Association National Test Station the first six years. The highly respected cattleman was selected to judge the World Congress when it was held in Calgary during the summer of 1982.

Clarence Ackert (pictured left), Kincardine, Ontario was CLA president in 1985. His wife Shirley served two terms as Limoselle president. The family has been especially active in the promotion of Ontario shows and sales. Ted Burnside (center), Guelph, Ontario might be very well tagged Mr. Performance Limousin. Ted has served on the CLA board and has been president of the Ontario association. He was chairman of the CLA performance committee for four years. Golden Limousin Acres, Calgary, Alberta is owned by John Dreher (pictured right), a past CLA board member. His efforts in breed promotion earned John and wife Marlene Limousin Leaders of the Year honors in 1985.

ing and yearling weights on the Canadian Sire Monitoring Program. Forty three head of bulls grossed $157,900 to average $3,680. Twenty-six fullblood bulls averaged $4,604 and the purebred and percentage bulls averaged $2,600 and $2,200 respectively. A record price in Canada was also established for the highest selling percentage bull at $3,850 selling to David Kenny of Rockyford, Alberta.

You can see by the end of the second year that the enthusiasm for the Canadian Limousin Bull Test Station had increased substantially because of the type and quality of the bulls and the prices received for top performance bulls.

The results of the Canadian Limousin Bull Test Station were beginning to be predictable, as the top indexing bull in 1980 was a bull called Arclan Legende who was a son of the 1978 bull test winner Limac Joe. A second son was sold as lot number 9 and a third son as lot 14. The bull sale in 1980 again was a success with the high selling bull, a son of Prince Pompadour, called Cedar Patch Dobi selling for $6,700 to Kevin Stanton of Seebe, Alberta and Lorne Bodell of Sherwood Parke, Alberta. Twenty-three fullblood bulls averaged $2,867 and six purebred bulls averaged $2,125 with 18 percentage bulls bringing $1,368. In all, 47 Limousin bulls sold for a gross of $103,325 with an average of $2,198.

The fourth Canadian Limousin Association Bull Test Station Sale was held on April 13, 1981. The top indexing bull was Del Mac Matador, a son of Goldenview Javelin consigned by Mackenzie Brothers and John Hormoth of Lethbridge, Alberta that sold for $13,500 to Bob Tabor of Delia, Alberta. The high selling bull "Big Mac" sold as lot 7 to LV Ranches and Berwyn Wise for $17,200. In total, 26 fullblood bulls sold for an average of $4,527, nine domestic purebred bulls averaged $3,056 and seven percentage bulls sold for an average of $1,850. In total, 42 lots grossed $158,150 for an average of $3,765.

The fifth Canadian Limousin Association Bull Test Station Sale was held on April 19, 1982. The top gaining and high selling bull was a son of Pub called Bonaparte Norseman consigned by the Bonaparte Ranch at Cache Creek, British Columbia. He sold to Glenkair Farms of Grandview, Manitoba for a new record price of $27,000. This is the bull that began a dynasty in the Canadian Limousin Bull Test Station in later years. There were 23 fullblood bulls which averaged $4,796 and eight purebreds averaged $2,194. A total of 31 lots grossed $127,850 for an average of $4,124.

In 1983 the Canadian Limousin Association Test Station was moved from the farm of Berwyn Wise at Irricana, Alberta to John Knight's farm located just 9 miles east of the Calgary city limits. At this time we established another record in that the high gaining fullblood bull in the 1983 sale was Bodell Rainmaker who sold to Ridge Ranch for $8,800. He was a son of Cedar Patch Dobi, who won the 1980 test, and was purchased by Lorne Bodell and Kevin Stanton for $6,700. One of the first sons of the Cedar Patch bull more than paid for his sire. Cedar Patch Dobi also had another son sell as lot number 6. Sons of previous test station high sellers and winners such as Limac Joe (from the 1978 sale), Mr. Eclaireur (from the 1979 sale) and Pub, the sire of Norseman (who won the 1982 station), were all dominant sires. In 1983 we tested a total of 133 head of bulls, the largest number to be tested in any one year.

In the 1983 sale there were 30 fullblood bulls which sold for $130,400 to average $4,346, 17 domestic purebreds grossed $35,650 to average $2,097 and six percentage bulls grossed $11,150 to average $1,858. In total, 53 head of bulls grossed $177,200 to average $3,343.

In 1984 we tested 130 bulls and we saw the beginning of a dynasty when GKF Canadian Ranger GKF 12R (the $27,000 Bonaparte Norseman son) was the top gaining bull at 3.66 pounds per day and sold to Cressman Cattle Company of Waterloo, Ontario and Mainline Limousin of London, Ontario for $24,500. The second high gaining bull was a son of Cedar Patch Dobi (the 1982 high seller) at 3.62

Curtis and Voris Molsberry, Leduc, Alberta and Lorne and Flossie Bodell, Sherwood Park, Alberta have held joint production sales for several years. Both firms have been active in the Alberta and CLA activities.

pounds per head per day. Pub, the sire of Norseman, had 12 out of 14 sons which were eligible for the 1984 test station sale. Eleven of his sons sold for a gross of $59,950 for an average of $5,450. We sold 40 fullblood bulls for a gross of $182,200 for an average of $4,550, 13 purebred bulls for $20,200 to average $1,554 and four percentage bulls for $5,700 to average $1,425. Fifty-seven head of bulls sold for a gross of $207,900 for an average of $3,647. This was the same year that Pub himself was sold to Stewart Farming of Big Valley, Alberta for $62,000.

In 1985 the high selling bull was GKF Canadian Showman (a son of Bonaparte Norseman, the 1981 winner) consigned by Cochrane Stock Farms, Alexander, Manitoba. This was the first time that a sire had produced a son which had won the test station two years in a row. The 1985 test station was the largest with 140 head of bulls tested. A high gaining bull, Hanchon's Napoleon (from the 1981 test), had sons standing second, fourth, eighth and eleventh. Sons of "Big Mac", Cedar Patch Dobi and Pub were also well represented. One-half interest in Canadian Showman sold to Bill Scriven of Ayton, Ontario for $14,900. There were 33½ fullblood bulls sold for $94,625 to average $2,825, six domestic purebreds grossed $9,925 to average $1,654 and one percentage bull sold for $2,150. The sale grossed $106,700 on 40½ lots to average $2,635.

In 1986 the high gaining bull out of the 128 tested was GKF Canadian Tripoly, a son of Bonaparte Norseman (the 1981 winner), with an average daily gain of 3.71 pounds per day. He was consigned by Glenkair Farms, Grandview, Manitoba and sold to Greenwood Limousin, Lloydminster, Saskatchewan for $8,250. Sixteen of the bulls that sold in the 1986 Bull Test Station Sale carried the influence of Pub or his sons and grandsons. Forty fullblood bulls sold for a gross of $127,925 for an average of $3,198, eight domestic purebred bulls grossed $21,625 for an average of $2,703 and two percentage bulls grossed $2,750 for an average of $1,375. In total, 50 lots grossed $152,300 for an average of $3,046.

In 1987 the high gaining bull was a son of the English imported bull Grahams Universal who was consigned by Hayard Farms of Innisfail, Alberta and sold to Belldoon Farms, Iona Station, Ontario for $4,500. In second place was a son of the Canadian Ranger bull (the 1984 winner) and in fifth place was a son of the Canadian Showman bull (the 1985 win-

Stan Cochrane, Alexander, Manitoba, 1986 Limousin Leader of the Year, was 1984 CLA president. The Cochrane Stock Farm herd has registered a number of significant successes including the top indexing and high selling bull in the 1984-85 National Bull Test. Pictured right is Gary Anderson, Bethune, Saskatchewan. He is currently serving as president of the Saskatchewan Limousin Association and serving on the CLA board of directors.

get the participation of at least five provinces and 40 to 50 consignors. Most persons observing the test program do not really understand how strenuous the selection program can be. Each year a maximum of forty to fifty bulls are selected from a total of 120 to 140 head. The chances of getting a bull to qualify in the top half are less than fifty percent and many breeders have tried for years to get a bull in the top half of the station. It is these standards that make it even more remarkable to have sons and grandsons of winning bulls come back and win the station. It has not only been one blood line but several individuals that have accomplished this feat. It is also interesting to note that all of the bulls that have been tested in the Canadian Limousin Association test program and been winners have also done well on the Sire Monitoring program.

The ultimate of course is to sell cattle for their full value and still provide the opportunity for the new owners to make money with their purchases. It is evident by our experience that the bull testing programs certainly help establish the genetic value and consistency in terms of performance. The style and the quality of the animal itself can be determined by the breeder. The confidence level and the price of animals with superior conformation and excellent

ner). In total, eighteen sons and or grandsons of the Pub bull were represented in the 1987 Bull Test Sale. One-half interest in a Harvest Olympus son for $7,250 proved to be the top selling bull which was an English imported bull consigned by Limousin Breeders of Canada, Sherwood Park, Alberta, selling to Molsberry Limousin of Leduc, Alberta. The high selling individual was once again a Canadian Ranger bull, Albertview Ulyses, selling to Combest Limousin of Stettler, Alberta. Forty-four and one-half fullblood bulls sold for a gross of $131,100 to average $2,944, six domestic purebreds grossed $11,450 to average $1,908. In total, 50½ lots grossed $142,450 to average $2,821.

The Canadian Limousin Association has been very successful in testing bulls. We have tested over 1,200 Limousin bulls and each year we have managed to

John Arnold, Waterville, Nova Scotia, has been a mover to establish the Maritime Association in the east with a membership today of over 30. He and Mark Cressman, Waterloo, Ontario were elected to the CLA board in 1985. Cressman was Ontario's first Limousin fieldman. The Cressman organization has been one of the industry leaders in the sale and show ring.

The 1987 CLA Board of Directors were elected at the annual meeting in Brandon. (Seated left to right) Gloria Antypowich, Limoselles representative, Horsefly, British Columbia; Mark Cressman, vice-president, Waterloo, Ontario; Gerry Good, president, Carstairs, Alberta; Wilbur Stewart, treasurer, Big Valley, Alberta; and Harvey Tedford, secretary-manager, Calgary, Alberta. (Back row from left to right) Robbie Garner, Simpson, Saskatchewan; Gary Anderson, Bethune, Saskatchewan; Yvan Brodeur, Contrecouers, Quebec; Bill Scriven, Ayton, Ontario; Henry Hays, Hardisty, Alberta; Bob Powell, Grandview, Manitoba; Clarence Ackert, Kincardine, Ontario; John Arnold, Waterville, Nova Scotia; and Lloyd Antypowich, Horsefly, British Columbia. (CLA Photo)

performance certainly rises dramatically. The price that breeders are willing to pay for performance tested yearling bulls when meaningful information is provided is truly remarkable. When you compare all of the other types of bull sales selling a volume of bulls and compare the averages of the bulls and the value buyers receive for the dollars invested, it is certainly the number one program.

The Canadian Limousin Association Bull Test Station has also succeeded because of the leadership that the committee chairmen have shown over the years and the maturity and realization by the members that their real bread and butter comes from production of sound performance cattle. You can compare this test station to all of the others in the country and you will find that other breeds cannot repeat our results. The reason is that this station is supported on a national basis by members from every province and the members themselves are actively involved in the day to day operation of the station and in the management of the station through the Bull Test Committee. The measurements taken at the station are not confined to rate of gain but also weight per day of age, testicle measurements and hip heights. All of these measurements are meaningful to various buyers and interested members and have been taken to supply important data to the members for use in their total bull evaluation. The Bull Test Station Committee has also been involved in evaluating the bulls for structural soundness and testicle size as criteria for the sale. This program has been received with mixed reactions over the years, but the success of individual breeders participating in the program has been greatly enhanced when all of the factors of selection were taken into consideration. I have taken some time to list the prices and the averages for all of the sales over the last ten years to illustrate the strength of the Bull Test Station program and point out to you that even though there are fluctuations of nearly 100 percent in the averages, these changes are not necessarily attributable to one particular factor but a variety of factors. Some of the changes were the result of the lack of participation by breeders in some years, the quality of the bulls in others combined with the general economic environment in the cattle business.

It has been interesting to note that despite all of these factors the Limousin Bull Test Program has been able to survive and flourish despite all of the normal growing pains and is a prominent, well respected institution that is supported by the membership as a source of bulls that are future breed herd sires.

There are other stations to test bulls in Canada; some are privately owned stations and others are government run stations. Many are operated in conjunction with university or college programs.

The rules and regulations are stringent and exacting, specifying rations, length of test, age, age spread, number of consignors in the test, number of sires represented and other economically important points in order to make results from one station to the next as uniform and meaningful as possible. Although the regulations, including health, are specified by the federal government, they are administered by the respective provinces.

Most facilities are all-breed stations and require prior planning to get bulls entered.

In Ontario a higher percentage of all bulls raised are tested because of that province's Red Meat Program. In this program a cattleman can collect $25 per head for commercial calves sold if sired by a performance tested bull that has indexed 100 or better. The program is self-policing in that breeders are fully aware that they eliminate potential buyers quickly without this stamp of approval.

Harvey Tedford thinks the trend for all over Canada is gaining momentum fast for performance testing—and especially Limousin. He cites the dramatic expansion of facilities by stations like the Regina station as documentation for his forecast in this growing trend.

Chapter 6
Conclusion

Without the foresight of the Canadian government to establish satisfactory quarantine stations in Canada, we could not be enjoying this Limousin breed as we are today. The trip with the breed the last 19 years has not been without its rough spots. In the late 70's, the disastrous cattle market saw many commercial herds dispersed, others cut herds back or reduced the heifer replacements they kept. The reductions increased the kill, putting further pressures on the dressed meat trade. The period reduced recording and registrations and the income for the association when it was badly needed.

The early eighties saw the market start to recover, then grind to a standstill for the association in 1984 and 1985 with a down cattle market, more cow herd reductions and a severe drought. In spite of these factors, demand was strong for Limousin bulls and prices were quite favorable when compared to other breeds in Canada.

As Harvey has mentioned in the manuscript earlier, it is dangerous to generalize, but I believe that it is an accepted fact that the Canadian breeders have had a higher priority in maintaining Limousin muscling as we saw it in France than the breeders of the United States have done. Perhaps the difference in meat grading standards in the two countries could be the underlying philosophy in creating the difference between the two countries.

One Canadian breeder, Don Matthews, commented on the subject to Debi Robbe in an interview, "Muscle is round, not flat. Judges talking about 'flat muscle' is for the birds." Many breeders on both sides of the border concur with the thinking because this is the carcass breed and it is the one breed that you can have total performance and muscle.

As the association and the breed moves into 1987, its future has never been brighter. Its breeders have taken a selected French foundation and tailored it well for Canada's commercial cattlemen.

TABLE TO REFERENCE SECTION

Leaders of Limousin History
in the United States

Founder Members

Robert H. Purdy, Buffalo, Wyoming	#1
Charles L. Moore, Des Moines, Iowa	#2
Bruce Waddle, Longmont, Colorado	#3
Sherm Ewing, Claresholm, Alberta, Canada	#4
James A. Scott, Bellvue, Colorado	#5
Mick Geesen, Agate, Colorado	#6
Stephen Garst, Coon Rapids, Iowa	#7
F. W. Gage, Denver, Colorado	#8
W. F. Morey, Loveland, Colorado	#9
Steve Lathrop, Wanamaker, Indiana	#10
H. A. McCoy, Seneca, Missouri	#11
Price/McCoy, Reading, Kansas	#12
The Garst Farm, Coon Rapids, Iowa	#13
Norris Cattle Company, Ocala, Florida	#14
Richard Goff, Denver, Colorado	#15
J. P. Woodbridge & Son, Pierson, Iowa	#16
Bob Evans, Bidwell, Ohio	#17
R. E. Shirbroun, Coon Rapids, Iowa	#18
Neil A. McKinnon, Bassano, Alberta, Canada	#19
Bov Import, Inc., Mead, Colorado	#20
National Bank of Waterloo, Waterloo, Iowa	#21
Dudley-Thompson Cattle Co., Hampton, Iowa	#21
John Airy, Johnston, Iowa	#23
Dr. Jorge de Alba, Coyoacan, Mexico	#24
Burwell Bates, Konawa, Oklahoma	#25
Margaret Shaw, Birch Tree, Missouri	#26
Thomas S. Christie, Billings, Montana	#27
Donald J. Kraus, Hays, Kansas	#28
Don A. Pavel, Longmont, Colorado	#29
Limousin Breeders of Canada, Calgary, Alberta, Canada	#30
Mildred L. Niehaus, Knoxville, Illinois	#31
Sam Hector, Calgary, Alberta, Canada	#32
J. Wright & L. Coleman, Melba, Iowa	#33
Karney J. Redman, Billings, Montana	#34
Sarah P. Forbes & Family, Sheridan, Wyoming	#35
Iven Blatter & Son, Malta, Montana	#36
Robert A. Beck, Bigfork, Montana	#37
Agri Cadre Ltd., Fort Collins, Colorado	#38
Roger D. Shaw, Birch Tree, Missouri	#39
Robert F. Crane, Jr., East Ridge, Tennessee	#40
Charles J. Wilson, Milwaukee, Wisconsin	#41
George C. Norris, Rush, Colorado	#42
Bradford Brothers, Richland Center, Wisconsin	#43
Arthur E. Rouse, Saratoga, Wyoming	#44
C.C. Burch, Johnstown, Colorado	#45
Richard K. Fenley, Louisville, Kentucky	#46
Robert C. de Baca, Huxley, Iowa	#47
J. C. Easland, De Smet, South Dakota	#48
John D. Moore, Newville, Pennsylvania	#49
J. J. Prosser, Mead, Colorado	#50
Frank G. Watson, Denver, Colorado	#51
George Demier, Miami, Oklahoma	#52
George Cocoros, Gaffney, South Carolina	#53
Leonard A. Coakley, Sterling, Colorado	#54
Nordic Farms, Campbellville, Ontario, Canada	#55
Robert Smellage, Boulder, Colorado	#56
Gerald Effertz, Velva, North Dakota	#57
Virginia & David R. Pulliam, Loveland, Colorado	#58
George Lampman, Toston, Montana	#59
Ron Baker & C&B Livestock, Hermiston, Oregon	#60
Joe Chase, Halliday, North Dakota	#61
Louis F. Swift, Fort Collins, Colorado	#62
Dundas Farms Ltd., Cardigan, Prince Edward Island, Canada	#63
Robert F. Baur, Van Meter, Iowa	#64
Louis de Neuville, Pueblo, Colorado	#65
Ed M. Arnold, Kilgore, Nebraska	#66
Mickey R. Collins, Winterburn, Alberta, Canada	#67
Lamont Brothers, Aberdeen, South Dakota	#68
Raymond Sutton & Son, Gettysburg, South Dakota	#69
Joe & Curtis Feist, Velva, North Dakota	#70
Steward & Morrissey, Inc., Baker, Oregon	#71
Bob Hartley, Vinita, Oklahoma	#72
Cadet Oxandaburu, Huron, South Dakota	#73
Bruce R. Burroughs, Oakley, California	#74
James L. Sutton, Onida, South Dakota	#75
W. J. Owen, Maxwell, Iowa	#76
John A. Doherty, Branson, Colorado	#77
Jack F. & Dorothy Doherty, Brady, Nebraska	#78
Ankony Angus Corporation, Grand Junction, Colorado	#79
Jose L. Trevino, Nuevo Leon, Mexico	#80
Phillip E. Flanagan, Riverton, Wyoming	#81
Robert Brooks, Tulsa, Oklahoma	#82
William Neely, Paris, Texas	#83
James E. Dula, Rubottom, Oklahoma	#84
Clarence Snyder, Clinton, Louisiana	#85
Jacques L. Leroy, Big Timber, Montana	#86
Jack McAuliffe, Redding, California	#87
Clayton Jennings, Highmore, South Dakota	#88
Ted Jennings, Highmore, South Dakota	#89
Harold Rinehart, Highmore, South Dakota	#90
Stuart Bush, Springfield, Missouri	#91
S. Neal Meyer, De Smet, South Dakota	#92
Bryant A. Harris, Garden City, Texas	#93
Hayes Mitchell Cattle Company, Marfa, Texas	#94
Frank Cable, Pukwana, South Dakota	#95
Frank Cable Jr. & Sons, Chamberlain, South Dakota	#96
John Zilverberg & Sons, Holabird, South Dakota	#97
J. L. Goodman, Texarkana, Texas	#98
John Frezieres, Grand Junction, Colorado	#99

Leaders of Limousin History
in the United States

Presidents

North American Limousin Foundation

1968-69	Bob Purdy	Buffalo, WY
1969-70	Bob Purdy	Buffalo, WY
1970-71	Bob Purdy	Buffalo, WY
1971-72	John D. Moore	Newville, PA
1972-73	Burwell Bates	Konawa, OK
1973-74	Bryant Harris	Marfa, TX
1974-75	J.V. "Lefty" Elliot	Fresno, CA
1975-76	Floyd McGown	San Antonio, TX
1976-77	Carlton Noyes	Minden, NE
1977-78	Carlton Noyes	Minden, NE
1978-79	William "Bill" S. Dameron	Salida, CO
1979-80	William "Bill" S. Dameron	Salida, CO
1980-81	Hal Courtney	Madill, OK
1981-82	Don Faidley	Colfax, IA
1982-83	Don Faidley	Colfax, IA
1983-84	Herman Symens	Amherst, SD
1984-85	Jim Davidson	Shepherd, MT
1985-86	Bob Yackley	Onida, SD
1986-87	Gene Raymond	Garnett, KS
1987-88	Leonard Wulf	Morris, MN

Executive Vice-Presidents

1968-Jan. 1972	Richard "Dick" Goff Denver, CO
Feb. 1972-Mar. 1972	Dale F. Runnion Fort Collins, CO
Apr. 1972-1977	Robert "Bob" H. Vantrease Denver, CO
1978 to present	Greg L. Martin Denver, Colorado

Board Members

Robert Purdy	Buffalo, WY	1968 to 1971
Sherman Ewing	Claresholm, Alberta, Canada	1968 to 1971 1974 to 1976
W. W. Smutz, Jr.	Denver, CO	1968, 1969
Charles L. Moore	Coon Rapids, IA	1968 to 1973
Stephen Garst	Coon Rapids, IA	1968 to 1973
H. A. McCoy	Topeka, KS	1968
Ben Price	Reading, KS	1968
Bruce Waddle	Longmont, CO	1968 to 1971 1984 to 1987
James A. Scott	Fort Collins, CO	1968 to 1970
John D. Moore	Newville, PA	1969 to 1973
Robert D. Bowers	Lake Francis, Manitoba, Canada	1969 to 1972
Burwell Bates	Konawa, OK	1969 to 1975
Waldo E. Forbes, Jr.	Sheridan, WY	1969 to 1971
James L. Baldridge	North Platte, NE	1970, 1971
Joe Chase	Halliday, ND	1970, 1971
J. C. Easland	De Smet, SD	1970 to 1975
Bryant Harris	Marfa, TX	1970 to 1976

Board Members

continued

Walter Shatto	Calgary, Alberta, Canada	1970 to 1973
Robert J. Steward	Keating, OR	1970 to 1973
David T. Allard	Bigfork, MT	1972, 1973
Fred DeMier	Miami, OK	1972, 1973
H. M. Jordan	Cleveland, MS	1972, 1973
Lee Leachman	Hotchkiss, CO	1972
Cadet Oxandaburu	Huron, SD	1972 to 1977
Don O'Brien	Pineville, MO	1973 to 1976
R. D. Bennett	Greenwood, FL	1973 to 1975
J. V. "Lefty" Elliott	Fresno, CA	1973 to 1976
Floyd McGown	San Antonio, TX	1973 to 1977 1981, 1982
William S. Dameron	Salida, CO	1974 to 1980
Carlton Noyes	Minden, NE	1974 to 1979
Hal H. Courtney	Madill, OK	1974 to 1980
Bryon N. Manley	Big Timber, MT	1974 to 1977
Truxton "Truck" Morrison	Godfrey, Illinois Wayzata, MN	1974, 1975 1977
William Schermer	Latimer, IA	1974 to 1977
Richard Dobson	Mannsville, OK	1976 to 1981
W. C. Tinsley	LaFayette, OK	1976 to 1981
William B. Worthington	Tucson, AZ	1976 to 1981
Leonard Wulf	Morris, MN	1976 to 1981 1985 to 1987
B. E. Greene	Vero Beach, FL	1977 to 1979
Daryl Nelsen	Maysville, MO	1977 to 1979
Ruth Price	Reading, KS	1977 to 1982
Ray Sims	Baird, TX	1978 to 1980
Bill Rumbley	Springville, CA	1978 to 1983
John Spivey	McDonough, GA	1978 to 1980
Herman Symens	Amherst, SD	1978 to 1984
Don Faidley	Colfax, IA	1978 to 1983
Ralph Copeland	North Platte, NE	1980 to 1982
Jim Davidson	Shepherd, MT	1980 to 1985
Alan Richardson	Clayton, IN	1980 to 1982
Darrell Wiggins	Houston, TX	1980
James Dyer	Davis, TX	1981 to 1987
Jerry Robbe	Pueblo, CO	1981 to 1984
Bob Yackley	Onida, SD	1981 to 1986
Peter Parker	Weed, CA	1982 to 1987
Gene Raymond	Garnett, KS	1982 to 1987
Fred Spitz	Okla. City, OK	1982 to 1984
Bob Williams	Decatur, Georgia	1982, 1983
James Fawley	Lynchburg, OH	1983 to 1987
Vernon Holcomb	Stanton, TX	1983 to 1987
Carl Johnson	Brandon, FL	1983 to 1987
Wyman Poe	Konawa, OK	1983, 1984
Jack Davis	Kalispell, MT	1984 to 1986
Ron Holland	Osage City, KS	1984 to 1987
John Williams	Atlanta, GA	1984, 1985
Bill Bridges	Miami, OK	1985 to 1987
Jerry Bradley	Braman, OK	1986, 1987
Bob Coscia	Springfield, MO	1986, 1987
Tom Grapner	Barnesville, GA	1986, 1987
Carl Crabtree	Grangeville, ID	1987
Rolf Mosbo	Rembrandt, IA	1987

Leaders of Limousin History
in the United States

Presidents
North American Limousin Junior Association

1971-72	Gloria Jennings	Highmore, SD
1972-73	Gloria Jennings	Highmore, SD
1973-74	Gloria Jennings	Highmore, SD
1974-75	Mark Leonard	Holstein, IA
1975-76	Mark Leonard	Holstein, IA
1976-77	Randy Bollum	Blue Earth, MN
1977-78	Randy Bollum	Blue Earth, MN
1978-79	Dick O'Brien	Pineville, MO
1979-80	Dick O'Brien	Pineville, MO
1980-81	Greg Bailey	Brownwood, TX
1981-82	Greg Bailey	Brownwood, TX
1982-83	Cheryl Linthicum	Welch, OK
1983-84	Cheryl Linthicum	Welch, OK
1984-85	Pete Anderson	Claremont, MN
1985-86	Kelly Long	Lynchburg, OH
1986-87	Kelly Long	Lynchburg, OH

Presidents
Limouselles—United States

1974-75	Alice Harris &	Marfa, TX
	Bonnie Booth	Welch, OK
1975-76	Bonnie Booth	Welch, OK
1976-77	Dee Kissler	Bennett, CO
1977-78	Dee Kissler	Bennett, CO
1978-80	Hariot Greene	Vero Beach, FL
1980-81	Janice Sappington	Greenwood, MO
1981-82	Janice Sappington	Greenwood, MO
1982-83	Dorothy Faidley	Colfax, IA
1983-84	Dorothy Faidley	Colfax, IA
1984-85	Pat Bingham	Justin, TX
1985-86	Eileen Bruner	Winfred, SD
1986-87	Eileen Bruner	Winfred, SD
1987-88	Rosalie Smith	Smithton, MO

Queens
North American Limousin Foundation

1971-72	Gloria Jennings	Highmore, SD
1972-73	Wyona Warren	Trinidad, CO
1973-74	Cindy Calavan	Warner, OK
1974-75	Kathy O'Brien	Pineville, MO
1975-76	Liz Crewson	Tulsa, OK
1976-77	Mary Susan Svoboda	Burbank, SD
1977-78	Theresa Scott	Rising Star, TX
1978-80	Lori Pihl	Marshall, MN
1980-81	Lori Leonard	Holstein, IA
1981-82	Cheryl Linthicum	Welch, OK
1982-83	Dee Jones	Sperry, OK
1983-84	Cindy Bollum	Blue Earth, MN
1984-85	Crystal Clark	Sumner, NE
1985-86	Renee Rupe	Edmond, OK
1986-87	Shannon Sewell	Eldorado Springs, MO
1987-88	Cinde Schuppe	Iliff, CO

Limouselle Scholarship Recipients

1978	Dee Jones	Sperry, OK
1979	Jennifer Eileen Platts	Culpepper, VA
1980	Card Sappington	Greenwood, MO
1981	Theresa Scott	Rising Star, TX
1982	Cheryl Linthicum	Welch, OK
1982	Thomas McDowra	Paytonville, TX
1983	Dana Sparkman	Miami, OK
1983	Mark Stovall	Hermosa, SD
1984	Wendy Droze	Crowder, OK
1984	Steve Kammerzell	Milliken, CO
1985	Linda Hankins	Tishomingo, OK
1985	Shana Ohman	Placerville, CA
1986	Renee Rupe	Edmond, OK
1986	Kristi Kent	Sweetwater, TX

Leaders of Limousin History
in Canada

Founder Members
Canadian Limousin Association

Limousin Breeders of Canada	Calgary, Alta.	#1000
Sam Hector	Calgary, Alta.	#1001
M. R. ''Mickey'' Collins	Winterburn, Alta.	#1002
L. J. Fouillard	St. Lazare, Man.	#1003
Sherm Ewing	Claresholm, Alta.	#1004
R. W. Tiffin	Gem, Alta.	#1005
S. M. Erwin	Calgary, Alta.	#1006
Joe Hochhausen	Edmonton, Alta.	#1007
Tony Rizzi	Enderby, B.C.	#1008
Bryce C. Stringham	Claresholm, Alta.	#1009
Ross I. Spence	Brooks, Alta.	#1010
Leslie Bain Hodges	Ponoka, Alta.	#1011
Romeo Laberge	Leoville, Sask.	#1012
William F. Hart	Midnapore, Alta.	#1013
William Perry	Caledon, Ont.	#1014
James W. Wallace	Westlock, Alta.	#1015
Berdo Investments Ltd.	Stettler, Alta.	#1016
Morris N. Palmer	Pincher Creek, Alta.	#1017
Dundas Farms	Cardigan, P.E.I.	#1018
John S. Wilson	Lacombe, Alta.	#1019
Louis Hochhausen	Edmonton, Alta.	#1020
Bridlebrook Farms Ltd.	Campbellville, Ont.	#1021
Lester C. Boulton	Okotoks, Alta.	#1022
Merle Derochie	Calgary, Alta.	#1023
Wilfred Hochhausen	Strome, Alta.	#1024
D. Bryon Palmer	Millarville, Alta.	#1025
Shatto Ranches	Calgary, Alta.	#1026
Dr. Walter Phillips	Calgary, Alta.	#1027
National Beef Breeders	Bassano, Alta.	#1028
Riverbend Limousin	Benalto, Alta.	#1032
Bov Import, Inc.	Denver, Colorado	#1034
Lake Francis Ranches	Lake Francis, Man.	#1041
C.L.R. Ranching Co.	Alsask, Sask.	#1043
David O. Randle	Calgary, Alta.	#1046
Earl P. Bergman	Erskine, Alta.	#1047
Victor Phillips	Beauvallon, Alta.	#1056
Don & Maurice Kirk	Dewinton, Alta.	#1059
Gordon Sherwood	Winfield, Alta.	#1063
A. J. Reidiger	Morden, Man.	#1077
Ray J. Nelson	Lloydminster, Alta.	#1086
Highland Stock Farms	Calgary, Alta.	#1087
F. E. Godwin	Millarville, Alta.	#1101
Norman T. McNally	Calgary, Alta.	#1105
D. R. Burke	Aldergrove, B.C.	#1133
Perrerin & Desaulniers	De Caxton Maurice, Que.	#1170
Jim McBride	High River, Alta.	#1180
Jim McLellan	Langley, B.C.	#1194
Nordic Farms	Campbellville, Ont.	#1202
Dennis Cassista	Cte Rimouski, Que.	#1367
John Lockhart	Okotoks, Alta.	#1600

Presidents
Canadian Limousin Association

1970	Joe Hochhausen	Strome, Alberta
1971	Joe Hochhausen	Strome, Alberta
1972	Walt Shatto	Calgary, Alberta
1973	F. E. ''Ted'' Godwin	Millarville, Alberta
1974	F. E. ''Ted'' Godwin	Millarville, Alberta
1975	Alan Parke	Cache Creek, British Columbia
1976	Alan Parke	Cache Creek, British Columbia
1977	Jim Lore	Carstairs, Alberta
1978	Jack Ward	Arrowwood, Alberta
1979	Dale Barclay	Erskine, Alberta
1980	Dale Barclay	Erskine, Alberta
1981	Don Matthews	Calgary, Alberta
1982	Mel Gosling	Dalemead, Alberta
1983	Mel Gosling	Dalemead, Alberta
1984	Stan Cochrane	Alexander, Manitoba
1985	Clarence Ackert	Kincardine, Ontario
1986	Gerry Good	Carstairs, Alberta
1987	Gerry Good	Carstairs, Alberta

Secretary-Managers
Canadian Limousin Association

1969-June 1972	Laura Palmer	Midnapore, Alberta
May 1972-1975	John Lockhart	Calgary, Alberta
1976 to present	Harvey Tedford	Calgary, Alberta

Presidents
Limoselles—Canada

1972	Gloria Holt	Calgary, Alberta
1973	Joyce McNally	Calgary, Alberta
1974	Joyce McNally	Calgary, Alberta
1975	Joyce McNally	Calgary, Alberta
1976	Mae Lockhart	Okotoks, Alberta
1977	Sandra Legge	Beechy, Saskatchewan
1978	Sandra Legge	Beechy, Saskatchewan
1979	June Lore	Carstairs, Alberta
1980	June Lore	Carstairs, Alberta
1981	Shirley Ackert	Kincardine, Ontario
1982	Shirley Ackert	Kincardine, Ontario
1983	Dorothy Sherwood	Winfield, Alberta
1984	Dorothy Sherwood	Winfield, Alberta
1985	Marg Sangster	Kenton, Manitoba
1986	Shirley Maurer	Windermere, British Columbia
1987	Gloria Antypowich	Horsefly, British Columbia

Leaders of Limousin History
in Canada

Board Members

Joe Hochhausen	Edmonton, Alta.	1969 to 1973
Walt Shatto	Calgary, Alta.	1969 to 1976
Sherm Ewing	Claresholm, Alta.	1969 to 1972
Stella Hart	Calgary, Alta.	1969
Byron Palmer	Midnapore, Alta.	1969 to 1973
Neil McKinnon	Bassano, Alta.	1969 to 1971
Mickey Collins	Winterburn, Alta.	1969, 1971 to 1973 1979 to 1981
William Perry	Caledon, Ont.	1970 to 1975
Bud McBride	Benalto, Alta.	1970 to 1973
Bryce Stringham	Lethbridge, Alta.	1970, 1971
Don Matthews	Calgary, Alta.	1970, 1971 1979 to 1983
Leo Fouillard	St. Lazare, Man.	1970
Paul Begin	Rimouski, Que.	1970
Ted Godwin	Midnapore, Alta.	1971 to 1974
Adrien de Moustier	Denver, CO	1972 to 1974
Stewart Erwin	Winnipeg, Man.	1972 to 1974
Romeo Laberge	Leoville, Sask.	1972, 1973
Norm McNally	Calgary, Alta.	1972 to 1975
Jim McLellan	Langley, B.C.	1972 to 1974
Don Morrison	High River, Alta.	1973 to 1975
Dennis Cluley	Campbellville, Ont.	1973, 1974
Alan Parke	Cache Creek, B.C.	1974 to 1980
Claire Denny	Alsask, Sask.	1974
Charlie Laird	Winnipeg, Man.	1974
Bill Kershaw	Calgary, Alta.	1974 to 1978
Ernie Shircliff	Sceptre, Sask.	1974, 1975
Stan Reimer	Marchand, Man.	1975 to 1977
Jim Lore	Carstairs, Alta.	1975 to 1978
Jim McBride	High River, Alta.	1975, 1976
M.R. "Bud" Boake	Acme, Alta.	1975 to 1977
Sandra Legge	Beechy, Sask.	1975 to 1979
Lloyd Atchison	Pipestone, Man.	1975, 1976
Lee Thomas	Stettler, Alta.	1975 to 1978
Jack Ward	Arrowwood, Alta.	1976 to 1980
Ted Burnside	Guelph, Ont.	1976 to 1980
Laurin Dreher	Weyburn, Sask.	1976 to 1978
Laura Palmer	Midnapore, Alta.	1976 to 1978
Dale Barclay	Erskine, Alta.	1977 to 1981
Ron Sangster	Kenton, Man.	1977 to 1979
Ian Outerbridge	Rockwood, Ont.	1977 to 1979
Stan Cochrane	Alexander, Man.	1978, 1980 to 1985
Gerry Long	London, Ont.	1978 to 1984
Keith McKinnon	Carseland, Alta.	1979 to 1984
Bill Karwandy	Cabri, Sask.	1979 to 1982
Clarence Ackert	Kincardine, Ont.	1980 to 1987
Les Allison	Gull Lake, Sask.	1980 to 1983
Mel Gosling	Dalemead, Alta.	1981 to 1986
Fritz Maurer	Windermere, B.C.	1981 to 1986
Don Downie	Blackie, Alta.	1981 to 1983
Roland Henuset	Pipestone, Man.	1982 to 1984
Jean Paul Lavoie	Rimouski, Que.	1982, 1983
Don Hauta	Dinsmore, Sask.	1983 to 1985
John Dreher	Calgary, Alta.	1983 to 1985
Bernard Payne	Lloydminster, Sask.	1984 to 1986
Gilles Lamoureux	Durvernay Laval, Que.	1984 to 1986
John Arnold	Waterville, N.S.	1985 to 1987
Mark Cressman	Waterloo, Ontario	1985 to 1987
Gerry Good	Carstairs, Alta.	1985 to 1987
Bill Scriven	Ayton, Ont.	1986, 1987
Wilbur Stewart	Big Valley, Alta.	1986, 1987
Robert Powell	Grandview, Man.	1986, 1987
Gary Anderson	Bethune, Sask.	1986, 1987
Robbie Garner	Simpson, Sask.	1987
Yvan Brodeur	Contrecouers, Que.	1987
Henry Hayes	Hardisty, Alta.	1987
Lloyd Antypowich	Horsefly, B.C.	1987

National Show Ring Champions

National Western Stock Show
Open Show—Champion Bulls

Year		Name	Sire	Exhibitor	Address
1979	Champion	Budrick's Innovator	Espoir de Carnaval	Budrick Farms	Mannsville, OK
	Reserve	Mastercharge	Inautiomare	Symens Bros. & Bain	Amherst, SD
1980	No Show				
1981	Champion	GKF Canadian Pacific 3L	Eclair	Glenkair Farms	Grandview, Man.
	Reserve	Black Ace	Dynamo	Black Jack Cattle Co.	Chattanooga, OK
				& Beaver Creek	& North Platte, NE
1982	Champion	Lookout Playboy (7/8)	Iris	Johannsen Limousin	Sutherland, IA
	Reserve	BLR Landmine (7/8)	BLR Landmark	Bates Limousin	Konawa, OK
1983	Champion	MK Sonic Jet	Seven Forty Seven	McKown Limousin &	Norman, OK
				Spitz Limousin Farms	
	Reserve	YKCC Atlantic (7/8)	Seven Forty Seven	Yackley-CMC Limousin	Onida, SD &
					McDonough, GA
1984	Champion	TEXS Ranger El Presidente	Seven Forty Seven	TEXS Limousin &	New Braunfels, TX
				JCL Cattle Co.	& Welch, OK
	Reserve	Quarterback	Espoir de Carnaval	Horizon Limousin	Madison, TX
1985	Champion	Spitz Navajo 607R	Seven Forty Seven	Spitz Limousin Farms	Norman, OK
	Reserve	TEXS Ranger Ted	MM Stretch	TEXS Ranger Ted Syndicate	Highmore, SD
1986	Champion	Spitz Navajo 607R	Seven Forty Seven	Spitz Limousin Farms	Norman, OK
	Reserve	Craftsman (purebred)	Dakota Klondike	Wulf Limousin Farms &	Morris, MN
				Cane Ridge Farms	Paris, KY
1987	Champion	Cressman Sky High	Roustan	Cressman Cattle Co. &	Waterloo, Ontario
				Taurus Breeders Service, Inc.	Lone Grove, OK
	Reserve	TEXS Ranger Premier	TEXS Ranger El Presidente	Premier Syndicate	Rosanky, TX

National Western Stock Show
Open Show—Champion Females

Year		Name	Sire	Exhibitor	Address
1979	Champion	SYBB Dakota Passion	Mr Sym Bros	Symens & Bain	Amherst, SD
	Reserve	SX Ms Sham's Eclair	Eclair	Card Sappington	Greenwood, MO
1980	No Show				
1981	Champion	CMC Miss Snowflake	Gibraltar	Yackley Limousin &	Onida, SD
				CMC Limousin	McDonough, GA
	Reserve	Miss Hermine 1M	Network	JCL Cattle Co. &	Welch, OK
				Hutchinson Farms	Thomson, GA
1982	Champion	YK Miss Enterprise	PURE Enterprise 777K	Hansen Limousin	Fort Collins, CO
		(Miss Denver)			
	Reserve	Metro Miss Galant 007M	Galant Guy CLR 4G	Jim Linthicum, Jr.	Welch, OK
1983	Champion	MK Lana	Seven Forty Seven	McKown and Baker	Norman, OK
	Reserve	HLF Kiss Me Quick	Espoir de Carnaval	Ellis-Vantrease Limousin	Blanchard, OK
1984	Champion	Dynasty (7/8)	WRC Punch	Yackley-CMC Limousin	Onida, SD &
					McDonough, GA
	Reserve	TEXS Star Gypsey	Seven Forty Seven	Tim Linthicum	Welch, OK
1985	Champion	Spitz Ara	Seven Forty Seven	Spitz Limousin Farms	Norman, OK
	Reserve	TEXS Star Luv	WRC Punch	James C. Linthicum	Welch, OK
1986	Champion	Spitz Special Effort	MK Sonic Jet	Spitz Limousin Farms	Norman, OK
	Reserve	MASH Surprise	Seven Forty Seven	Horizon Limousin	Madisonville, TX
1987	Champion	LML American Dream	Dr. Ray	Little Mondeaux Limousin	Genoa, NV
	Reserve	TEXS Star Genie	TEXS Ranger El Presidente	Joli Limousin Farm &	Bellevue, WA &
				TEXS Limousin/	New Braunfels, TX/
				JCL Cattle Co.	Welch, OK

National Western Stock Show
Champion Pens and Carloads

Year	Champion Pen Grand	Champion Pen Reserve	Champion Carload Grand	Champion Carload Reserve
1987	Symens Brothers Amherst, SD (yearlings)	YK-CMC-LML Onida, SD (calves)	Leonard Wulf and Sons Morris, MN (yearlings)	YK-CMC-LML Onida, SD (calves)
1986	YK-CMC-LML (yearlings)	Symens Brothers (calves)	YK-CMC-LML (calves)	Leonard Wulf and Sons (calves)
1985	Yackley-CMC (yearlings)	Symens Brothers (yearlings)	Yackley-CMC (calves)	Symens Brothers (calves)
1984	Symens Brothers (calves)	Spitz Farms Norman, OK (yearlings)	Yackley-CMC (calves)	Symens Brothers (calves)
1983	Symens Bros. and Bain (twos)	Peterson's L7 Bar Limousin Pukwana, SD (calves)	Yackley-CMC (calves)	Lookout Limousin Timber Lake, SD (calves)
1982	Symens Brothers (yearlings)	Yackley-CMC (calves)	Yackley-CMC (yearlings)	Lookout Limousin (yearlings)
1981	Lookout Limousin (yearlings)	Yackley-CMC (calves)	Lookout Limousin (calves)	Pompadour Hills Highmore, SD (calves)
1980	Symens Brothers (yearlings)	Ox & W Huron, SD (yearlings)	Lookout Limousin (calves)	Dameron Land & Cattle Salida, CO (yearlings)
1979	Deux Amis Limousin Rising Star, TX (calves)	Symens Brothers (2-year-olds)	William Schermer & Sons Latimer, IA (yearlings)	Dameron Land & Cattle (2-year-olds)
1978	Yackley Limousin Ranch Onida, SD	Bates Limousin Ranch Konawa, OK	Dameron Land & Cattle (2-year-olds)	Pompadour Hills (yearlings)
1977	Yackley Limousin (calves)	Adams Limousin Ranch (calves)	Dameron Land & Cattle (yearlings)	Pompadour Hills (yearlings)
1976	Adams Limousin Ranch (calves)	Peterson's L7 Bar Limousin (calves)	Pompadour Hills (calves)	Coble Ranch Seminole, OK (yearlings)
1975	Yackley Limousin	Follis Limousin Ranch Miami, OK (yearlings)	Dameron Land & Cattle (calves)	

Canadian Western Agribition
Open Show—Champion Bulls

Year		Name	Sire	Exhibitor	Address
1974	Champion	Lacombe Elite	Dandin-C	A.L.E. Farms	St. Lazare, Man.
	Reserve	K & D Beau Fleck KDS 1F	Famous	K & D Farms	Wawota, Sask.
1975	Champion	Fanfaron's Revolution 10F	Fanfaron	Laurent & Richard Maguet	St. Rose du Lac, Man.
	Reserve	Highland Gendarme 104G	Gendarme	Highland Stock Farms	Calgary, Alta.
1976	Champion	Highland Gendarme 94G	Gendarme	Highland Stock Farms	Calgary, Alta.
	Reserve	Goldenview Hugo GVC 29H	Fox Trot	Goldenview Farms	High River, Alta.
1977	Champion	Glenkair Fanfaron 69H	Fanfaron	Glenkair Farms	Grandview, Man.
	Reserve	Highland Gonzague Blaise SJL 2H	Gonzague	Highland Stock Farms & Jeanne Locke	Calgary, Alta. & Stettler, Alta.
1978	Champion	Highland Ideal CMA 99J	Ideal	Highland Stock Farms	Calgary, Alta.
	Reserve	Glenkair El Toro GKF 7K	El Toro	Glenkair Farms	Grandview, Man.
1979	Champion	Goldenview Kolossal	Goldenview Hercules	Goldenview Farms	High River, Alta.
	Reserve	Golden Limo Legend	Goldenview Hercules	Golden Limousin Acres	Calgary, Alta.
1980	Champion	Glenkair Great Event 75L	Famous	Glenkair Farms	Grandview, Man.
	Reserve	GKF Canadian Pacific 3L	Eclair	Glenkair Farms	Grandview, Man.
1981	Champion	Glenkair Great Event 75L	Famous	Glenkair Farms	Grandview, Man.
	Reserve	Roustan (Import)	Major	Golden Limousin Acres	Calgary, Alta.
1982	Champion	Sir Limolyn CLF 055M	Sir Echo	Combest Limousin	Erskine, Alta.
	Reserve	Bonaparte Norseman	Pub	Glenkair Farms	Grandview, Man.
1983	Champion	Bodell Northstar BLR 1N	Eager	Guth Brothers	Unity, Sask.
	Reserve	Golden Limo Regis	Roustan	Golden Limousin Acres	Calgary, Alta.
1984	Champion	Sir Limolyn Regent CLF 306R	GKF Canadian Pacific	Combest Limousin	Erskine, Alta.
	Reserve	Greenwood Shogun 85S	MM Stretch	Greenwood Limousin	Lloydminster, Sask.
1985	Champion	Goldenview Royal GVC 1R	Goldenview Laredo	Highland Stock Farms	Calgary, Alta.
	Reserve	Today's Request	Seven Forty Seven	Greenwood Limousin	Lloydminster, Sask.
1986	Champion	Golden Limo Ulternator	Captain In Command	John Dreher	Calgary, Alta.
	Reserve	Today's Request	Seven Forty Seven	Greenwood Limousin	Lloydminster, Sask.

Canadian Western Agribition
Open Show—Champion Females

Year		Name	Sire	Exhibitor	Address
1974	Champion	K & D Belle Gazelle (import)	Cesar 01	K & D Farms	Wawota, Sask.
	Reserve	Fanfaron Modern Lady 3F	Fanfaron	Laurent Maguet	Ste. Rose du Lac, Man.
1975	Champion	Guincheuse (import)	Espoir	Goldenview Farms	High River, Alta.
	Reserve	Goldenview Lady MCB 4F	Diese	McBride Limousin	High River, Alta.
1976	Champion	Ivresse (import)	Danseur	McBride Limousin	High River, Alta.
	Reserve	Goldenview Lady Diese GVC 3F	Diese	Goldenview Farms	High River, Alta.
1977	Champion	Highland Henrietta CMA 97H	Gendarme	Highland Stock Farms	Calgary, Alta.
	Reserve	Miss Glenkair 67G	Fanfaron	Glenkair Farms	Grandview, Man.
1978	Champion	Prairielane Immobile	Cherokee	MRS Limousin	Kenton, Man.
	Reserve	Miss Gonz El Toro	Gonzague Blaise	JRL Limousin	Stettler, Alta.
1979	Champion	Goldenview Lady	Goldenview Hercules	Goldenview Farms	High River, Alta.
	Reserve	K & D Belle Jasper	Gonzague Blaise	K & D Limousin	Wawota, Sask.
1980	Champion	Highland Henrietta CMA 97H	Gendarme	Highland Stock Farms	Calgary, Alta.
	Reserve	Miss Circle T EMT 70L	Eclair	Circle T Limousin	Outram, Sask.
1981	Champion	Manitoba Nahane GKF 13N	GKF Canadian Pacific	Glenkair Farms	Grandview, Man.
	Reserve	Michelle MRS 1M	Deruset Jasper	MRS Limousin	Kenton, Man.
1982	Champion	Saskatchewan Lily PYN 36N	Helix	Guth Brothers	Unity, Sask.
	Reserve	Manitoba Crocus II	Glenkair Famous	Glenkair Farms	Grandview, Man.
1983	Champion	Highland Pixie CMA 105P	GKF Canadian Pacific	Highland Stock Farms	Calgary, Alta.
	Reserve	Highland Ransom CMA 44R	Highland Conversation CMA 111N	Highland Stock Farms	Calgary, Alta.
1984	Champion	Golden Limo Rosette	Roustan	Spence W. Pepper	Didsbury, Alta.
	Reserve	Manitoba Sophia GKF 25S	Bonaparte Norseman	Glenkair Farms	Grandview, Man.
1985	Champion	Manitoba Sophia GKF 25S	Bonaparte Norseman	Blue Sky Limousin	Donalda, Alta.
	Reserve	Limolyn Conversation CLF 335R	Highland Conversation 109N	Combest Limousin	Erskine, Alta.
1986	Champion	Manitoba Tanya GKF 66P	Bonaparte Norseman	Glenkair Farms	Grandview, Man.
	Reserve	Knolland Trillium	Helix	Knolland Farms	Norwood, Ont.

National Junior Heifer Show Champions

Year		Name	Sire	Exhibitor	Address
1976	Grand	DANL 511G	Eros	Dan Tetrick	Redwood Falls, MN
	Reserve	Miss Cheryl 121G	Eclair	Cheryl Mewes	Highmore, SD
1977	Grand	YKCC Desert Queen	Famous	Steve Yackley	Onida, SD
	Reserve	MH Gentle Thoughts	Eclair	Debbie Jackson	Miami, OK
1978	Grand	''Yeller''	Fanfaron	Vernon McKown	Norman, OK
	Reserve	Miss Favori 377J	Prince Favori	Cheryl Linthicum	Welch, OK
1979	Grand	Miss Cheryl 81K	Carnaval	Cheryl Linthicum	Welch, OK
	Reserve	821K Lookout Nita	Eclair	Julie Carmichael	Timber Lake, SD
1980	Grand	Miss Guy 565L	Galant Guy	James Linthicum, Jr.	Welch, OK
	Reserve	CMC Miss Snowflake	Gibraltar	Steve Yackley	Onida, SD
1981	Champion	Genuine Risk	MM Stretch	Dan Mewes	Highmore, SD
	Reserve	YKCC Miss Jackpot	Jackpot	Steve Yackley	Onida, SD
1982	Champion	TEXS Star 878N	Seven Forty Seven	Neil Scott	Rising Star, TX
	Reserve	MK Lana	Seven Forty Seven	Vernon McKown	Norman, OK
1983	Champion	MK Jennifer	Seven Forty Seven	Richard McKown	Norman, OK
	Reserve	TEXS Star Gypsey	Seven Forty Seven	Timmy Linthicum	Welch, OK
1984	Champion	MK Tiffany	Seven Forty Seven	Richard McKown	Norman, OK
	Reserve	TEXS Star Luv	Punch	James Linthicum, Jr.	Welch, OK
1985	Champion	MK Stephanie	Snowball	Richard McKown	Norman, OK
	Reserve	Lace	Atlantic	Suzanne Stewman	Maryneal, TX
1986	Champion	Sweet 'N Sassy	Captain In Command	Shari Holloway	Chattanooga, OK
	Reserve	TEXS Star Genie	TEXS Ranger El Presidente	Timmy Linthicum	Welch, OK

Chronology of Breed History

*Significant events in the development of the
Limousin breed in North America.

1968

May 6—First Pre-Organizational meeting held in Denver.
Purdy, Goff, C. Moore, Waddle, Scott, Smutz, Patterson, Carpenter, de Moustier present. Founder member
plan adopted. Ad hoc committee appointed to proceed
with association organization plans, Goff, Purdy,
Smutz.

June 20—Foundation incorporated as a non-profit organization in the state of Colorado.

June 21—Articles of Incorporation and By-Laws adopted.
Board elected: Purdy; W. W. Smutz, Denver, Colorado;
Steve Garst, Coon Rapids, Iowa; H. A. McCoy, Topeka,
Kansas; Chas. Moore, Coon Rapids, Iowa; Ben Price,
Jr., Reading, Kansas; James A. Scott, Fort Collins, Colorado; Bruce Waddle, Longmont, Colorado. Purdy
elected president; Sherm Ewing, Claresholm, Alberta, vice president; Smutz, secretary-treasurer. Goff
named Executive Vice President.

July 1—Castor (Prince Pompadour) becomes first
Limousin in North America.

Aug. 8—Semen contract signed for Prince Pompadour
semen.

1969

July 1—The "D" bulls arrive. Bov's Diplomate and Dandy;
ABS's Decor and Curtis' Doremi.

July 25—Board asks Dick Goff to take on operation of
NALF on a full time basis. Negotiations were successful.

Nov.—Prefix system adopted. Performance program
discussion continues.

1970

Mar.—The grand inaugural meeting of the Canadian
Limousin Association is held at the Palliser Hotel in
Calgary, Alberta.

April—Top performance bull in Montana BIAS Test Station purchased from Bob Purdy for carcass cut-out research. Twelve steers purchased from Mick Geesen
for cut-out data.

First issue of the Canadian Association Newsletter
is published.

Canadian Limousin Association (CLA) organized and
reported in NALF board meeting to have 300 members.

July—U.S. quarantine station being considered by USDA.

NALF sends $500 to CLA to help defray organizational
expenses.

Dec.—Standard terms and conditions for National Sale
adopted. Jim Baldridge, North Platte, Nebraska to
manage January '71 sale at National Western in
Denver.

1971

Jan.—First National Sale held in Denver. Dudley Thompson halfblood tops sale at $10,000. 103 head average
$1,437. The 13 yearling halfbloods averaged $3,752.

May—Volume 1 Number 1 of the *International Limousin
Journal* is published.

First and second state Limousin associations formed
are Oklahoma and Texas. Other groups in the process of organizing include South Dakota and Nebraska.

A southeastern association is in the process of forming to include Florida, Georgia, Alabama and Mississippi. Washington recently organized, with Kansas,
Missouri and Colorado having scheduled meetings for
this purpose. Intermountain will include states of Colorado, Wyoming, New Mexico and Utah.

Garst Farms first production sale in Coon Rapids,
Iowa saw 20 halfblood heifers average $2,765. All were
bred to Limousin.

Black Champ Farm, Waxahachie, Texas schedules
first Limousin sale in that state.

Feature of the Red Carpet Sale, Huron, South Dakota
was the first Limousin queen contest in North America
with Gloria Jennings, Highmore, South Dakota winning the title.

Dean Jacobs, North Platte, Nebraska joins *Journal*
staff as field editor.

Espoir de Carnaval (by Carnaval bred by Beyrand)
comes on line producing semen in the Bov Import, Inc.
bull battery.

Aug.—Five Limousin steers sired by Dandy and entered
by Bov Import, Inc., Denver, Colorado won championship honors in feedlot gain test division of recent
Great Western Beef Expo, Sterling, Colorado.

Limousin crossbred steers take grand and reserve at
Calgary Stampede.

Sept.—Iowa forms state Limousin association bringing
to eight the number of states or regions that have
organized.

Formation of American Cattle Services as a management and consultant service for breeding was announced August 1 by Ken Holloway, Frederick,
Oklahoma.

Oct.—Over 650 visitors from 11 states and two foreign
countries attend Oxandaburu Limousin "Show and
Tell" Field Day at the ranch, Huron, South Dakota.

Nov. 30—Oklahoma Limousin Association holds first
show and sale of Limousin in U.S. Bates Limousin,
Konawa, Oklahoma exhibits first champion. A 50%
Limousin by Prince Pompadour tops sale at $5,000.
45 lots average $1,394.

* most dates used in this section are the dates when the events
were published in the official breed magazine in the United
States.

Dec. 6—First 75% heifer to sell at auction by Baldco Farms, North Platte, Nebraska to Buster and Irene Jones, Jacksonville, Illinois for $10,500.

1972

Jan. 10—Goff resigns as executive vice president of NALF.

Jan. 14—Two fullblood bulls, Computer and Challenger, sell in National sale in Denver for $56,000 and $40,000 respectively. (First fullbloods to come to the states.)

Feb.—Dale F. Runnion appointed executive vice president of NALF.

Mar.—Darrel Menning, Corsica, South Dakota became the North American Limousin Foundation's 1,000th member.

North American Limousin Foundation reaches membership list of 1,101—101 new members so far in 1972.

Junior members worked hard at the 1972 National Western selling buttons. They now have $225 in the bank.

Geesen Ranches sold a 75% blood female in the Southwest National sale for $9,900.

Dandin C (by Noel from Chastanet) a 4-year-old Limousin bull (Fullblood) set a Canadian record at Calgary when he sold to International Beef Breeders Inc. of Denver for $176,000.

Six fullblood bulls sell in the Canadian Legacy sale to average $51,167.

Jerry Adamson, Cody, Nebraska sets all time price record in carload feeder division of the National Western Stock Show with first Limousin-Angus cross-bred steers. They sold for an average of $581 per head and were sired by Diplomate and Dandy.

April—Keystone Commander comes to U.S.—first generation fullblood bulls (also Comet) purchased in Canada for export to U.S.

First 75% bull calf to sell at auction (consigned by Road Runner Farms, Clinton, LA) brought $1,000.

Twenty-two breeders meeting in Reno, Nevada form five state (Washington, Oregon, California, Idaho, Nevada) Limousin Associaton. J.V. Elliott, Fresno, California president.

Vantrease named executive vice president of NALF. Has Bachelor of Science in Animal Science from University of Tennessee. Was member of 1962 International Livestock judging team.

June—Memberships are up to 1320, that's 320 since December 31. By June 1st should be 20,000 females on performance program—by June 1st some percentage female or bull will be assigned registration #15,000.

Limousin cross steers will vie for $10,000 cash and championship honors in Limousin Carcass show to be called Limousin Bonanza.

A Limousin-Shorthorn cross steer was Supreme Grand Champion at Western Stock Show, Edmonton, Alberta, Canada, shown by Lynn Armistead, Onaway, Alberta. Outstanding carcass also.

31 fullblood heifers and bulls arrive in Calgary, bring total for 1972 French-born Limousin to 217.

Organization of Australian Limousin Breeders Society with Jock Makin as president.

July—Mickey Collins calved the first fullblood Limousin by embryo transplant at his Flying Red Wheel Ranch, Winterburn, Alberta.

Aug.—Limousin sire Dandy sets three records at 1972 Great Western Beef Expo: more pounds of live steer per day of age, more carcass weight per day of age and more pounds of quality red meat per day of age. All weights were Expo records.

Membership is now up to 1,400.

New name adopted for the Limousin Performance Program is BEEF which means Breeding Efficiently with Essential Facts.

Limousin-Hereford crossbred steer takes championship at the Calgary Stampede—second year in a row.

75% Limousin bull topped 140 day gain test at the Southwest Sire Evaluation Center, Frederick, OK. Gained 4.68 per day on official test and weighed 1,195 lb. at completion.

Sept.—NALF adopts sire evaluation program and will proceed in developing the first sire summary in next few months.

Registration requirements changed: bulls of 94% Limousin blood and females of 87% are accepted for registration in the Limousin registry book providing performance data is submitted. (Purebred defined.)

NALF members honor Dick Goff and Dale Runnion for outstanding service to the foundation—plaques presented by John Moore and Bob Vantrease into the "Order of the Red Head."

Oct.—Over 25 champions from county, district or state fairs reported to NALF office during the summer.

BEEF program in operation, and it is now possible to print pedigrees from program.

110 Limousin-Angus cross steers slaughtered in August at St. Joseph, Missouri after being hauled over 1,000 miles. Weight was 1,085 when killed, carcass weight 704 lbs. 71% choice, 56% yield grade 2's and 43% yield grade 3's. Excellent set of steers that prove Limousin is the carcass breed.

Grand Champion steer at Ozark Empire Fair was Limousin exhibited by Keith Kissee. Limousin also won beef carcass contest and exotic cross bred heifer competition in open class.

Nov.—1,000 Limousin cross steers slaughtered this summer and data sent to NALF. 70% graded choice, 90% yield grade 1, 2, or 3.

Jack Eberspacher of Beaver Crossing, Nebraska won the grand championship of the country's largest steer show, Ak-Sar-Ben, with Limousin cross steer.

Limousin carcasses took both championship honors at second annual Golden Pride Carcass show in Sioux Falls, South Dakota.

40 Americans from 8 states tour Paris and Limousin country in Limoges.

Dec.—American Limousin Herd Book is established. All Limousin of 50% blood or better will be recorded in book. Will replace previous two herd book system.

Limousin steers swept 1972 Royal Winter Fair Crossbred show: Scored 1-2-3 in lightweight steers, 1-2-3-4 and 5 in middleweight class of 27 steers. Also grand champion and reserve grand champion honors.

1973

Jan.—Limousin won over 50 steer shows in 1972.

Dragueuse, a fullblood imported female from Pompadour Estates herd in France sold safe in calf to Esca for $21,500 in the first Saskatchewan sale to BST Ranch at Estevan. She was the first French Limousin import to sell at public auction.

Limousin cross steers made a clean sweep at the 1972 International Livestock Exposition, Chicago Amphitheatre. Grand champion pen of five over all breeds bred and shown by Hollars Ranch, Cody, Nebraska and reserve grand champion pen of feeders was bred and shown by Rocking J Limousin Ranch, Cody, Nebraska. Both were Limousin.

77 fullblood bulls now on the North American continent. Chart published in *International Limousin Journal* lists all by breeder name.

Feb.—Limousin progress at National Western: 3 out of 4 classes in steer show, stood 1-2-3 in one class; 1st and 2nd in two other classes. Limousin won champion and reserve champion in crossbred division and reserve grand champion of the show. In carload division Limousin were only exotic breed to show and 3 of 4 crossbred loads were Limousin. Grand champion over all breeds show was Triple L Limousin, Hotchkiss, Colorado.

Memberships and recordations on rapid increase. In one year membership increased by 850 and in last four months (Sept., Oct., Nov. and Dec.) recorded 10,000 cattle.

Arkansas to have meeting to organize association in February.

Mar.—34,000 cattle recorded and 1,900 members. 3,000 cattle were recorded in January along with 98 new members.

Southwestern Select Limousin Sale, Fort Worth, Texas featured the sale of the breed's first pure French Limousin female, Harbinger for $50,000, selling to Rick and Sharon Dobson, Mannsville, Oklahoma. Harbinger was imported from England.

Tom Gaskell, Denver, Colorado joins NALF staff as Communications Director.

Laura Palmer is awarded the first honorary membership of the Canadian Limousin Association.

Apr.—March 14 was the day 2,000th member joined the NALF. There are now 16 purebred bulls in the U.S. and 39,000 cattle recorded.

First 87% bull calf (by Echo) was born recently in the Limousin herd of H. M. Jordan, Cleveland, Mississippi.

Arkansas breeders organize Limousin association. Stuart Bush, president.

Six Alberta breeders buy 650 halfblood Limousin heifers bred for three-quarter calves from Tex Fulton, Miller, South Dakota.

May—March was largest month to date for both memberships and recordings. 107 new members and 4,200 cattle recorded.

Bob Crook, editor and advertising manager of *Cattle Guard* magazine, joins the staff of *International Limousin Journal*.

First Sire Summary appears in the *International Limousin Journal*.

CLA reports having 1,181 memberships and has registered 6,900 Limousin.

Canada's first fullblood out of imported stock was born at Bov Import, Rimouski, Quebec (bull calf Fantastic by Dandy out of Danseuse).

June—Registrations double in first 5 months of 1973— now at 60,000.

July—Dr. Larry Benyshek has been appointed Director of Research and Education for the North American Limousin Foundation.

Limousin-Angus cross steer took Supreme Championship honors at the Great Western Junior Livestock show in Los Angeles, California. He was also carcass champion.

Aug.—Limousin sweep crossbred steer show at Calgary Exhibition for 3rd consecutive year, grand champion, reserve grand champion, best pen of 3 and inter-club group class.

Sept.—French Connection sale held on lawn of Regency Inn, Denver, Colorado. 8 pure French Limousin (3 bulls, 5 females). First opportunity for U.S. buyers to bid on pure French mature Limousin bulls that were born in France.

Limousin Carcass Bonanza held July 26-28 in Denver, daring new show paying $14,604 in prizes. Judged on basis of performance and carcass merit with more than 80 steer entries and 29 exhibitors from nine states. Retail value of grand champion value per day was $2.03, weighed 1290 lbs. (459 days of age), and had weight per day of age at 2.81 lbs. 1.36 yield grade with ribeye at 18.4 sq. in., grading good.

1973 Illinois State Fair grand champion steer was a Limousin crossbred shown by James Horsley, Toulon. Steer sold for $7500.

Oct.—Limousin have won more than 70 steer shows so far this year.

Inaugural meeting of the International Limousin Council held in Limoges, France Sept. 11-17 with delegates from 20 countries. Louis de Neuville elected president.

Grand champion of exotic breeds at Ozark Empire Fair was a heifer shown by Keith Kissee, Ozark, Missouri.

Limousin win 8 out of top 10 places at Colorado State Fair Beef carcass contest. First 5 are all Limousin cross steers.

Membership up to 2,470 and registrations 78,000.

Mr LCCO establishes a breed record yearling weight at 1,335 lbs. Official adj. wt. is 1,440 lbs. (highest actual ylg. wt. and adj. ylg. wt. recorded by NALF).

Nov.—Jerry Adamson wins feeder show at American Royal, Eastern National, Cow Palace with Limousin.

Dec.—Kim Anderson, Ohio, Illinois and Val Eberspacher, Beaver Crossing, Nebraska, showed the grand champion and reserve grand champions respectively at the 1973 American Royal in Kansas City. Limousin steers also took champion and reserve champion honors in FFA division champion in 4-H division and grand champion pen of 5 feeder steers.

1973 grand champion steer at Grand National Livestock Exposition was Limousin cross shown by Dan Kemp.

1974

Jan.—Title of movie produced by NALF is "7000 Years New." Narrated by Ben Johnson, movie actor, Academy Award winner and Limousin breeder.

Announcement is made that Sire Summary completed by January 1974 and will include birth, weaning and yearling information (more than 75,000 records used to evaluate data).

Grand champion carcass at State Fair of Texas (Dallas) was Limousin-Angus crossbred owned by Linda Hancock, Midland, Texas.

Feb.—69 lots average $6,796 in the 4th National Sale held in Denver during National Western Stock Show.

Toughest competition ever in National Western Stock Show Pen Feeder Steers gives grand championship honors to Rocking J Limousin, Cody, Nebraska. It was the fifth major show grand championship to be captured by Jerry Adamson feeder steers.

First issue of *Limousin Leader,* Canada's Limousin breed publication, is printed and delivered to breeders attending 1974 CLA convention in Calgary, Alberta.

Mar.—Limousin steer nabs grand champion honors at Southwest Exhibition and Fat Stock Show in Fort Worth.

Top honors at San Antonio Livestock Exposition went to three outstanding Limousin steers: grand champion on foot, reserve grand champion on foot and grand champion carcass.

First World Limousin Futurity is announced for September 26-Oct. 1 in Tulsa, Oklahoma.

Meeting in Fort Wayne, Indiana saw formation of Indiana-Michigan Limousin Association.

North American Sire List shows seven fullblood bulls from the United Kingdom now in U.S.

Complete list of pure French Limousin sires in North America is published.

Apr.—Kirby and Lillian Briggs, Kirley, South Dakota are owners of the 100,000 registration in the NALF Herd Book.

Fourteen fullbloods sell in ballroom sale of the Calgary Inn to average $19,321.

May—Bull studs report their biggest year selling Limousin semen.

Dean Jacobs, advertising manager of *International Limousin Journal* resigns post to become sale manager headquartered in Nebraska.

June—Southeastern Limousin Association Sale in Atlanta brings average of $4,585.

Eastern Limousin Association is formed at John Moore's farm in Newville, Pennsylvania.

Dan Wedman, Yukon, Oklahoma and graduate of Oklahoma State University joins staff of *International Limousin Journal.*

Sale of 31 Limousin from Sweden, all purebreds, is announced by Raymond Hefner, Oklahoma City, Oklahoma. Purchaser was Woody Sudbrink, Top Hat Ranch, Madison, Florida.

Aug.—Eclair sired group of steers wins Supreme championship at Bonanza Field Day. Owned by Kirby and Lillian Briggs, of Midland, South Dakota.

Limousin steer wins grand championship for 4th consecutive year in on foot competition at Calgary Exhibition and Stampede.

Sept.—Limousin cross steers are grand champion at Iowa, Illinois and Ozark Empire Fairs. Second time in succession for Limousin at Iowa and Ozark Empire.

Oct.—Enthusiastic Limousin breeders meet in Versailles, Kentucky to organize state association.

Registrations for year number 34,075 and transfers have reached 19,087. Total registrations 110,000. 609 new members during last year for a total of 3,150 members in 40 states.

Nov.—Limousin Ladies organization (Limouselles) is formed at convention in Tulsa.

First place in Ak-Sar-Ben Livestock Show 4-H beef carcass contest is Limousin steer.

First World Limousin Futurity in Tulsa, Oklahoma brings 180 entries in show and sale averages $3,812 on 82 lots.

"Greatest ever" convention in Tulsa, speakers Dr. Robert Totusek, Martin Jorgenson, Dr. Bernard Jones and Dr. Lowell Wilson focus on breed efficiency. Kathy O'Brien new national queen.

Dec.—San Francisco's Grand National grand champion steer is Limousin cross.

1975

Jan.—North Dakota Limousin breeders organize state association with Martin Weekes, Raleigh as president.

First National Western Stock Show where percentage Limousin bulls had a show classification in carload and pens show.

Value on Limousin crosses was emphasized at 1974 National Western auction sale of pens of feeders in the yards with ten pens of Limousin crosses averaging $452.45. High selling pen was sired by 75% Limousin bulls of Jerry Hollers, Cody, Nebraska.

Apr.—Dameron sells 80 high percentage bulls to Hilton Ranch, Nevada.

June—English expert on embryo transplant, Dr. Jim Bowen is interviewed by *ILJ.* He is currently visiting professor at CSU, has fellowship in reproduction from Cambridge.

First 15/16 bull calf is born at Dr. and Mrs. Richard Shirbroun in Coon Rapids, Iowa. This is considered a purebred.

July—Michelle Cockram, Baker, Oregon shows grand champion crossbred steer at 1974 Pacific International Exposition.

Aug.—Fifth consecutive year in a row that Limousin cross is grand champion in live steer competition at the Calgary Exhibition and Stampede.

Sept.—September 15—Limousin breeders of Tennessee will meet to organize state association. September 17—Limousin breeders of Georgia will meet to organize state association.

Nov.—Mark Leonard, Holstein, Iowa is re-elected president of the North American Limousin Junior Association.

Worlds largest 4-H show falls to Limousin steers with grand championship honors taken by Chris Maier, Eagle Grove, Iowa. Reserve spot went to J. D. Anderson, Arapahoe, Nebraska, also with a Limousin.

Limousin crossbred carcass entries take John Morrell competition at Sioux Falls, South Dakota for second straight year.

Dec.—As of November 1 there were 160 (100%) purebred females and 200 (100%) purebred bulls registered in the U.S.

Illinois Limousin Association formed with Bob Robertson, Rockford, as president.

Limousin steer calves were grand champion feeder pen at 1975 American Royal. Shown by Jerry Adamson.

1976

Jan.—First American Invitational Limousin Show and Sale held November 17, 1975 in Louisville, Kentucky with 17 consignors from 10 states exhibiting 38 head.

Feb.—14 breeders from 5 states exhibit pens of three bulls and 3 breeders exhibited carloads during the second year Limousin have been a part of the National Western Carload and Pen Show.

Dee Kissler, Bennett, Colorado, organized a Limouselle breakfast and style show during the National Western.

First Limousin Leader of the Year award, sponsored by the *Limousin Leader* is awarded to Jim & Ruth McBride, High River, Alberta and Christina Baumann Massie, Campbellville, Ontario.

May—Grand champion steer at the 45th annual Wichita Falls District Beef Show in Texas was a 1,160 lb. Limousin-Angus crossbred shown by Libbi Garnett, Vernon.

Robert H. Purdy, Buffalo, Wyoming rancher and first president of the North American Limousin Foundation died April 1. He was 59.

June—Craig Schrader, Indiana native and 1976 graduate of Purdue University accepted a post as field editor with the *International Limousin Journal.*

Nancy Arlene Hoffman and Dan Wedman, field editor for the *International Limousin Journal*, wed July 17 in Yukon, Oklahoma.

Sept.—A 75% Carnaval son was named grand champion steer at 1976 Open To The World Steer Show held in conjunction with the Iowa State Fair.

IBS, Inc. of Seminole, Oklahoma averages $6,400 on 27 fullblood lots.

Iowa holds Open to World junior heifer show in Des Moines, Iowa on August 20. 44 exhibitors from 6 states attend. Grand champion heifer honors went to Dan Tetrick, Redwood, Minnesota.

Oct.—First Limousin fullblood sale in United States to feature All-Irish breeding, August 22 in Osceola, Iowa averages $3,933 on 23 lots.

Nov.—MH Gibraltar sells to syndicate for $55,000—October 16, 1976.

Limousin crossbred steer win at Ak-Sar-Ben, world's largest 4-H steer competition second year in a row.

Bov Import purchases International Beef Breeders.

Recently wed in Oklahoma were NALF Executive Vice President Bob Vantrease and Mary Griswold.

1977

Feb.—92 pens shown in National Western Feeder Pen Show with Larry Hollers capturing champion crossbred and grand champion pen with Limousin crosses.

Apr.—Limousin steers get off to great start in 1977 by sweeping grand champion honors at Denver, Fort Worth, San Antonio and Houston. All this less than ten years after arrival of the breed in the U.S.

Aug.—Andy Rest, Colorado State University graduate appointed editorial assistant after *Limousin Journal* internship.

Sept.—First National Junior Heifer Show is held at Springfield, Missouri in conjunction with Ozark Empire Fair.

Bob Vantrease resigns as executive vice president of NALF.

Dick Goff given special recognition from French government for his part in introducing the Limousin breed in North America.

Floyd McGown (TEXS), San Antonio, Texas elected president International Limousin Council.

1978

Feb.—Eighth National Limousin Sale with standing room only crowd averages $3,888.

Pen and carload show in Denver grows in numbers and quality with 4 carloads and 23 pens in competition. Demand for Limousin bulls and prices paid were talk of the yards. 58½ bulls average $2,756 at private treaty.

Mar.—Two year old study highly favorable to Limousin. Ease of calving and carcass traits highly rated.

South Dakota Red Carpet sale beats 1977 sale by $400.

Apr.—Brahmousin name selected for Brahman-Limousin blood cattle.

May—Holcomb sisters, Staunton, Texas show both grand and reserve grand Limousin champions at Houston.

June—Non-surgical technique lowers cost of embryo transplants. Keeping large numbers of available recipients a major cost.

Seven-eighths Limousin born in Costa Rica from frozen embryo shipped from Canada.

July—*International Limousin Journal* publishes 402 page Herd Reference Issue as breed strength grows in U.S. and Canada.

Aug.—Brahmousin guidelines for registering purebred and percentage Brahmousin established.

Sept.—NALF observes 10th anniversary of Foundation.

Nov.—Promotion piece "The Word" (the word is Limousin) published by NALF.

Dec.—Dee Jones, Sperry, Oklahoma awarded first Limouselle scholarship.

San Francisco Grand National adds Limousin to show schedule.

Dameron sells 77 percentage bulls in Nebraska Sand Hills sale averaging $1,475. Top is $3,700.

1979

Jan.—Deux Amis Futurity Sale, November 19, 1978 featured 77 fullblood transplant calves out of 7 females and 5 bulls averaging $3,914.

First Limousin show held in conjunction with the Pacific International Livestock Exposition, in Portland, Oregon. 44 head from 3 states were shown.

One of the foundation herds of the breed, "PRO" Easland Cattle Company dispersed December 11 & 12, 1978 in Huron, South Dakota. 431 lots sold to 21 states for a strong $2,007 average.

For the third year in a row a Limousin steer won the grand champion carcass award at the Ag-Ex in Brandon, Manitoba.

Colorado Junior Association was formed in December, 1978.

One of the breed's first double polled animals to sell at auction was a 3/25/78 purebred son of Polled King, selling November 16, 1978.

Feb.—Highest price ever paid for an entire carload of bulls purchased by 1 buyer in the 73 year history of the National Western Stock Show in Denver was $35,000 for the 14 members of the reserve grand champion carload of Limousin bulls owned by Dameron Land and Cattle Co. of Salida, Colorado. The load sold to Grand River Ranch of Kremmling, Colorado.

Polled Northern, a polled purebred son of Espoir owned by Canadian Polled Limousin Enterprises of Alberta sold for $12,000 for one-half interest to Leland Dudley of Hampton, Iowa.

New record for a percentage bull was set in 1978 when a seven-eighths son of Carnaval, YK Treetop, sold for $9,000 at the Yackley Limousin Bull Sale to Leon Tucker of Milford, Indiana.

Goldenview Farms of High River, Alberta had the top selling bull in 1978 as they sold one-half interest in Goldenview Hercules to McKown & Sons Limousin, Ardmore, Oklahoma in a private treaty transaction for $60,000.

Top selling purebred honors, at auction in 1978, went to Riverbend Honneur RBF 71H. Wiggins Ranch Company of Cleveland, Texas sold one-half interest in the Eclair son to Rancho Santa Clara of Santa Clara, Mexico for $30,000.

Richard & Sally Knight, Pierce City, Missouri paid $30,000 for Jericho, a purebred son of Famous at the Yackley Limousin Ranch bull sale in Onida, South Dakota.

Top selling female honors in 1978 went to Godwin Golden Helen BTR 06 and Modern Lady. Both females brought a final bid of $17,000.

One-half interest in Visa, a black 15/16 Carnaval son sold to Jim Meyer, Wagner, SD in 1978 for a record $50,000. The William Schermer Family of Latimer, Iowa raised the bull.

The first time in history of the Limousin breed at the National Western Stock Show in Denver a black purebred was named reserve champion heifer. She was exhibited by Yackley Limousin Ranch, Onida, South Dakota.

One-half interest in Black Ace, a 15/16 black purebred bull sold to Beaver Creek Limousin, North Platte, Nebraska and Loti, Kansas for $10,000. The line bred Castor was bred by Black Jack Cattle Company of Chattanooga, OK.

National Western grand champion pen of three—exhibited by Deux Amis Limousin, Rising Star, Texas. National Western reserve grand champion pen of three—exhibited by Symens Brothers, Amherst, South Dakota. National Western grand champion carload—exhibited by William Schermer & Sons, Latimer, Iowa. National Western reserve grand champion carload—exhibited by Dameron Land & Cattle Company, Salida, Colorado.

Six pens of 41 feeder calf pens at the Arizona National Livestock Show were Limousin crossbreds. Brocker Sage Creek Ranch, Walden, Colorado had the first place middleweight and lightweight pens and sold them for $1,974 average.

A 1,263 lb. three-eighths Limousin steer, exhibited by Troy Thomas, Harold, South Dakota was named grand champion steer at the 1979 National Western Stock Show in Denver, Colorado.

Mar.—Grand champion steer at the Fort Worth Stock Show was a five-eighths black Limousin steer exhibited by Sonya Deatherage, Stanton, Texas. Topping the 600 head show, this steer went on to sell for a record $52,000.

Reserve grand champion steer at the Fort Worth Stock Show was a halfblood black Limousin steer owned by Becky Linsay, Mullin, Texas. Seven of the top 12 crossbred steers carried Limousin blood.

On Feb. 9, 1979 the highest price ever paid for a bull of any breed in the history of the Black Hills Stock Show in Rapid City, South Dakota was $7,100 for a two-year-old percentage son of Mr. Symbros, consigned by Symens Brothers, Amherst, South Dakota.

A Pompadour Hills bred halfblood Limousin steer was named grand champion at the San Antonio Livestock Exposition.

Apr.—Limousin cross steers took first place honors in every major junior show in Texas this year: grand and reserve—Fort Worth; grand and reserve—El Paso; grand and reserve—Amarillo; grand and reserve—Abilene Stock Show; grand and reserve—San Angelo; grand—San Antonio; grand—Houston; grand—Wichita Falls; grand champion carcass—Fort Worth and San Antonio; grand—Oklahoma City Junior Livestock Show.

New world record set when grand champion steer at Houston Livestock Show sells for $70,000. Jim Bob Beam, Patrick, Texas exhibited the seven-eighths Carnaval son bred in the Yackley Limousin herd.

Andrew M. Rest, *Limousin Journal* editorial assistant for the past two years has been named the *International Limousin Journal's* field editor for Northwest and Western United States.

Bob Vantrease, former executive vice-president of NALF and Dr. Robert Ellis of Oklahoma City form new breeding partnership, Ellis-Vantrease, based in Amber, Oklahoma.

Goldenwest Farms, Miami, Oklahoma exhibited the grand champion purebred bull at the Houston Fat Stock Show and Rodeo for the third year in a row.

Sudbrink Percentage Limousin herd is dispersed. 213 lots sold to 47 buyers from 7 states.

Mar. 8, 1979—1st Kansas Beef Expo takes place in Wichita, Kansas.

May—Apr. 7—The first purebred and percentage Brahmousin are offered at auction at the Wiggins 1st Texas "Blockbuster" Sale, Marble Falls, Texas.

Apr. 7—Nordic Debutante brings $55,000 at Wiggins Sale, selling to Lew Bingham of Aubrey, Texas.

North American Limouselle Association will continue to provide an annual $200 college scholarship to qualifying junior members.

The third National Junior Heifer show will be held in Stillwater, Oklahoma at the conclusion of the Limousin Breeders Symposium, July 12-14 on the campus of Oklahoma State University.

North American Limousin Foundation office began receiving Brahmousin applications for recordation.

June—May 11 & 12—The world's largest ever offering of Full French Limousin sells at the DeMier-Sudbrink Sale in Miami, Oklahoma. This sale becomes the breed's first million dollar event. 323 2/3 lots gross $1,659,000 for a $5,128 average.

Hanchon, senior herd sire of the Sudbrink program paced the million dollar sale by bringing $61,000 for a two-thirds interest.

Apr. 24—YK Jackpot, a black purebred son of Gibraltar sells for $34,000 at the Yackley Limousin Bull Sale, Onida, South Dakota to a group of leading Limousin breeders.

Apr. 16—Second Canadian Limousin Association Bull Test sale has top selling bull bring $15,200. Top selling percentage bull brings $3,850.

Rio Vista Farms, San Antonio, Texas offers planned matings through their transplant operation at a cost of $3,000 to $4,000 with a minimum of four ova purchased. Cost of transplanting, for each safe pregnancy, is $1,800.

The first calf ever born of a frozen embryo exported from the United States was born June 1978 in Germany.

The first U.S. calf to come from a frozen embryo was born in Feb. 1978.

Aug.—The first Annual American Royal National Limousin Show will be held in Kansas City, Missouri, Oct. 29.

The first Colorado Limousin Junior Field Day was held June 16 in Loveland, Colorado.

July 12-14—First Limousin Breeders Symposium, "Crossroads '79" was held in Stillwater, Oklahoma.

Sept.—Dale F. and June A. Runnion, founders and publishers of *International Limousin Journal* since 1971 have announced the sale of the magazine to Bob Crook, a partner since 1977.

Cattle Management Consultants, a sale management firm at North Platte, Nebraska have changed their name to Cattle Brokers, Inc. and have hired Milt Sexson, formerly a field editor for the *International Limousin Journal.*

Oct.—Aug. 25—Fleuralie sells to H.A. McCoy, Seneca, Missouri for $31,000 at Wiggins Ranch Co. "Fabulous Factory Sale."

Las Vegas, Nevada—Grand and reserve grand steers were Limousin cross.

August 10—Grand champion honors at the Berkley County Youth Fair and Baby Beef Show & Sale in Martingsburg, West Virginia went to a 66% Limousin steer.

For the first time a Limousin steer won the Ohio State Fair in Columbus. He sold for $20,000.

Nov.—Ak-Sar-Ben Livestock Exposition has 697 entries. Reserve grand champion is a Limousin steer. Limousin steers were named champions in three divisions and reserve in two divisions.

Sept. 1979—First ever Limousin show is held in conjunction with the Waterloo Iowa Congress.

Dec.—Oct. 29—One of the largest open Limousin shows ever held in North America happened at the first American Royal National Limousin Show.

Grand champion steer at the Tulsa State Fair is a Limousin cross.

1980

Jan.—Nov. 16—The 8th Annual Farm Fair "Pacesetter" Show and Sale held in Edmonton, Alberta had 235 entries making it the largest Limousin show ever held.

The top four steers at Edmonton's Northlands Farm Fair were Limousin cross steers.

Grand champion and reserve grand champion market steers at the 1979 Canadian Western Agribition in Regina, Sask. were Limousin-Angus crosses.

Nov. 19—Limousin Breeders of Canada disperse after ten years of solid selection and management.

Nov.—The first Limousin show held at the Royal Agricultural Winter Fair in Toronto, Ontario has 104 head.

In 1980 Canadian Limousin breeders will import 69 head of fullblood Limousin; 30 head will end up at Goldenview Farms of High River, Alberta.

Nov. 30—At the Polled Limousin Dispersal Sale of Woody Sudbrink Ranch, Miami, Oklahoma, Jethro, a purebred Carnaval son, sells for $31,500.

Four new members are added to the *International Limousin Journal* staff: Carroll Cannon will be field editor for mideastern United States and portions of Canada; Phil Wilgers will be business manager and pilot; Bob Collins is hired as graphic artist; Jill Wood will be office manager and assistant editor.

Dec. 10 & 11, 1979—Largest performance-oriented program, Dameron Land and Cattle, dispersed in Platte City, Missouri. 800 head sold to 90 buyers from 20 states in the second Limousin sale to gross over 1 million dollars. All but two head were three-quarter American purebreds.

Grand champion carcass steer at Louisville, Kentucky was a black Limousin cross steer that had a 17.8 inch rib eye and graded choice with a 65.6 dressing percentage.

Grand champion steer at the Kansas Inter-State Fair was a Limousin cross.

Dec. 15—One of the oldest Limousin herds in the nation, Bell-Wood Limousin of Osceola, Iowa disperses.

Nov. 4—Mar del Plata, Argentina was the site of the 4th International Limousin Council meeting presided over by President Floyd McGown of TEXS Limousin, San Antonio, Texas. Peter Cullen of England was elected the new president.

Minnesota 4-H Market Livestock Show, Jackson, Minnesota had 200 head entered. A three-quarter Limousin steer won the overall competition which included rate-of-gain, carcass quality and live evaluation. Four of the top ten were Limousin.

Feb.—The largest private treaty sale in Limousin breed history has taken place between Bob Hutchison and Sudbrink Limousin. The sale included 280 fullblood Limousin for $2,000,000.

Miss Cheryl 81K, grand champion female at the 1980 National Limousin Show and Sale sells for $31,000.

SYBB Dakota Slick, the grand champion bull sold for $30,000 at the 1980 National Show and Sale.

Rio Vista International announces that custom embryo freezing will be available to the cattle industry during the 1980's.

Merv Mewes, Highmore, South Dakota announced the sale of one-tenth interest in MM Stretch to TEXS Limousin, San Antonio, Texas for an undisclosed price.

Apr.—The 1980 grand champion steer at the Houston Stock Show and Rodeo was a Limousin crossbred that sold for $110,000.

The Limousin breed has its first show at the Southern Arizona International Livestock Association Show.

Limousin sale at the Black Hills Stock Show in Rapid City, South Dakota out averaged all other breeds.

May—Apr. 5—At Wiggins "V.I.P." Sale, Nordic Debonnaire sells for $33,000. In two years Debonnaire has produced 159 embryos with 110 confirmed pregnancies.

A three-quarter black Limousin steer was named grand champion steer over 650 entries at the Oklahoma Junior Livestock Show in Oklahoma City and sold for $20,640.

June—Apr. 19—Amazon, a black, polled, purebred bull sold for $30,000 at the Smith Cattle Company "Black Power" sale. Five yearling black purebred bulls averaged $11,080.

A three-quarter Limousin cross steer swept the competition at the 1980 San Antonio Livestock Exposition and set a new San Antonio record of $27,500 at auction.

July/Aug.—Elijah, a fullblood son of Eli out of the Joye cow in the Ellis-Vantrease herd set a new Limousin record for weaning weight of a fullblood bull. Actual 205 day weight was 880 lb.

May 29—One-sixth interest and 400 vials of semen in YK Jackpot, a black purebred son of Gibraltar brings $10,000 at the Sutton-Yackley Ranch Dispersal, Onida, South Dakota. Jackpot's mother, an 87 percent black female sold for $16,500.

June 14—The first Texas Junior Limousin Association meeting is held in conjunction with the Texas Junior Heifer Show.

Hillbright Farms, Marietta, Georgia purchased the entier Auld Clootie full French Limousin herd in Caledon, Ontario.

May 27—Christina Baumann Massie of Nordic Farms presents a fullblood bull to Mexican President Joe Lopez Portillo at a reception hosted by the Mexican Embassy in Ottawa, Ontario. The bull, Toron, will be used for crossbreeding purposes at Portillo's cattle ranch as well as in the A.I. beef industry in Mexico.

Sept.—Carroll Cannon, *International Limousin Journal* field editor married Patsie Tompkins, July 20, 1980. Cannon will now be representing the ILJ as the eastern field representative.

Leonard Wulf, Morris, Minnesota was recently named the Minnesota Commercial Cattleman of the Year.

The first Brahmousin certificate was issued.

Nov.—Percentage Limousin steers claimed both the grand and reserve grand champion banners at the 723 head Ak-Sar-Ben Livestock Exposition.

Louisiana Limousin Association held their organizational meeting in the summer of 1980.

Dec.—Keith Kissee of Ozark, Missouri will join the Worrell Land and Cattle Company in Charlottesville, Virginia in Dec. as manager of the Worrell Limousin herd.

Champion steer carcass of 27 breeds at the Lafayette County, Missouri 4-H Livestock Show was an 87 percent Limousin steer with a 66.6 dressing percentage.

A Limousin cross steer topped 130 entries of all breeds at the Iowa Performance Carcass Classic with a yield grade of .6 and a rib eye of 18.1 in.

1981

Jan.—Dec. 10—Molsberry Limousin Ranch 1st Production Sale, 17 lots average $10,453.

Highland Henrietta 97H has become the first Limousin female to win all of the major Canadian shows in one year.

Ten fullblood and purebred pairs averaged over $7,300 at the 4th Annual Bell Ringer Consignment Sale, Calgary, Alberta.

Limousin entries came away with five top awards at the Northlands Farm Fair steer show in Edmonton, Alberta, winning grand champion steer, champion and reserve junior steer and champion and reserve pen of three.

The largest single breed performance test in the United States in 1981 is at La Junta, Colorado where 266 Limousin bulls from 12 Colorado breeders are on test.

Feb.—Jan. 15—The National Limousin Sale during the National Western Stock Show in Denver was held for the first time in the Beef Palace Sales arena "on the hill."

The grand champion female, CMC Miss Crystal, sells for $30,000.

Jill Wood, an employee of the *International Limousin Journal* (ILJ) since Mar. 1979 is named to the ILJ editor's post. She becomes the first distaff editor of the publication in its 10 year history.

Jan. 16—The first ever Open Limousin Show in the 75 year history of the National Western Stock Show attracted 120 head from 15 states and 4 Canadian provinces.

CMC Miss Snowflake and GKF Canadian Pacific take champion honors at the first Limousin Open show at the National Western Stock Show in Denver, Colorado.

Grand champion carcass at the North American Livestock Exposition was a Limousin-Angus cross.

Jan. 6—First Limousin show is held in conjunction with the Arizona National Livestock Exposition, Phoenix, Arizona.

Lookout Limousin, Timber Lake, South Dakota has champion carload at the National Western for the second year in a row. They also win champion pen of three honors.

Mar.—Jan. 27—Dudley-Thompson Limousin disperses one of the most productive cow herds in North America on the site of the first Limousin sale ever held in the United States.

Jan. 29—Ray Sims, past director of the North American Limousin Foundation and one of the most widely known and highly respected livestock breeders in the country passed away after a 10 month battle with cancer.

CMC Miss Snowflake 102L becomes the Limousin breeds first $102,000 female as two-thirds interest sells for $68,000.

Feb. 6—Grand champion steer at the 1981 Southwest Livestock Exposition, Fort Worth, Texas was a seven-eighths Limousin steer that sold for $32,000 in the premium sale. A seventy-five percent black Limousin steer was the reserve champion.

Grand and reserve grand champion carcass at the 1981 Arizona National Livestock Show went to two Limousin-Angus cross steers.

Apr.—Mar. 9—A new breed record for a percentage bull is set when Mr. Symbros 66L, "Spot," sells for $12,000 at Symens Brothers first production sale, Amherst, South Dakota. Of the 78 bulls offered not one sold under $2,000. SYBB Dakota Lass 49L sells for $23,000.

Mar. 12—Leland Dudley, owner of Dudley-Thompson Cattle Company, the first Limousin operation to hold a production sale in the United States, passed away after a battle with cancer.

Mar. 8—Eclair, a foundation sire of the Limousin breed celebrates his twelfth birthday and the production of more than 60,000 recorded progeny.

Feb. 21—Manitoba Crocus sells for $37,500 and GKF Canadian Pacific, the 1981 National Champion, brings $26,000 at Glenkair Farms second production sale in Brandon, Manitoba.

May—Apr. 3 & 4—Wiggins Ranch Company Complete Dispersal in Marble Falls, Texas shatters Limousin record books. Highest sale gross—$2,408,565—5,721 average. Most dollars sold at auction (of Limousin cattle) in one day—$1,610,240 (Apr. 3...251 lots average $6,415). Most Limousin females sold for $10,000 or more in one day—24 (Apr. 3). Most Limousin lots sold for $10,000 or more in a single sale—34. Highest gross on a full sibling pair in one sale—$112,000.

Napoleon and WRC Deb's Lady both sired by Gendarme and out of Debonnaire sell for $77,000 and $35,000 respectively.

Canadian Limousin Association Bull Sale record is set as LV Ranches, Stettler, Alberta and Wise Ranches, Irricana, Alberta teamed up to pay $17,200 for the Hayard Farms consignment, "Big Mac."

Topping the William Schermer and Sons annual bull sale was a double polled purebred son of WS Jethro that sold for $15,000 to Hayard Farms, Innisfail, Alberta, Canada.

For the first time in Carcass Breed history a full French Limousin bull born in the United States was sold back across the border to Canada. The son of Carnaval out of Hillary was bred by H.A. McCoy and sold to Dr. Keith Wakefield, Red Deer, Alberta.

192

June—YKCC Cadillac Jack sets a new American record for a black, purebred calf when he sold to Jim London, Freedom, Oklahoma for $36,000.

May 16—49 lots in the North Texas Limousin Classic sale held near Justin, Texas average $6,328. Top selling female of the sale was MH Gentle Thoughts who sold to Worrell Land and Cattle Company for $39,500.

Kodak, a 3/24/80 son of Northern Dancer and Something Happened, owned by the Ellis-Vantrease Limousin Ranch, Blanchard, Oklahoma was recently syndicated for $100,000.

The purchase price of $12,500 for a black, three-quarter son of Mastercharge represents a new record for the private treaty sale of a percentage bull. Kalvin and Dale Tupps, Byers, Colorado sold the bull to three Texas cattlemen.

McKown and Sons Limousin Ranch, Norman and Ardmore, Oklahoma have announced the recent one-half interest sale in Seven Forty Seven to Worrell Land and Cattle Company, Charlottesville, Virginia for a new record price in the Limousin breed of $150,000. The previous record was $176,000 for full interest in Dandin-C in 1972.

May 4—The "Fabulous Factory" Sale held at Nordic Farms in Campbellville, Ontario averaged $8,180 on 94 lots. 27 Full French pairs averaged $11,835. Top selling female was Nordic Jouita, selling to Worrell Land & Cattle Company.

A three-quarter Limousin steer was named reserve grand champion at the most prestigious steer show in the world, the Houston Fat Stock Show and Rodeo.

July/Aug.—May 23—Hutchinson Limousin Farms' "C'est Magnifique" sale averaged $6,761 on 83 1/10 lots. Miss Hermine 1M topped the sale by selling to Bingham Limousin Ranch, Justin, Texas and McKown and Sons Limousin Ranch, Ardmore & Norman, Oklahoma for $70,000, a new high for a Limousin female at auction.

Bates Limousin Ranch, Konowa, Oklahoma has announced eight purebred Limousin heifers and one fullblood bull from the Bates program have been shipped to the Republic of Panama to be used by the Panamanian government. This is the first shipment of Limousin breeding stock from the U.S. to any Central or South American country.

TEXS Limousin Ranch, New Braunfels, Texas has announced the sale of 14 bulls to three cattle ranchers in northern Mexico.

Sept.—Terry Jaschke has been hired as managing editor of the *International Limousin Journal*.

A new record for a black, purebred bull was set when the final semen interest in "Black I" sold for $10,700 for a final tally of $80,700.

Oct.—Limousin breeder, Bob Hutchinson, 32, of Thomson, Georgia was killed September 2, in a plane crash which also claimed the lives of his two brothers, James and Thomas.

Sept. 5—Fancy Britches BTR 09F, a half-sister to Daffodil and 747 sold for $52,000 in the Nease & Son Limousin dispersal in Oklahoma.

One-sixth interest in YK Jackpot, a black, purebred son of Gibraltar sold for $7,800 at the Bar T Ranch Dispersal near Lockhart, Texas.

Nov.—Bates Limousin Ranch of Konawa, Oklahoma has announced the sale of one-half interest in BLR Kavalier to New Breeds Industries of Manhattan, Kansas for $20,000.

Dec.—One of Texas' leading Limousin breeders, Richard Charles Howe died October 19 of a heart attack suffered at his Clear Creek Ranch home near Forestburg, Texas.

1982

Jan.—Limousin steers totally dominated every carcass show in Canada in 1981. Brandon Ag-Ex—grand and reserve grand champion, top 7 steers were Limousin. Calgary Round-Up—grand champion, 6 of top 8 steers were Limousin. Edmonton Farm Fair—grand and reserve grand champion, top 5 steers were Limousin. Regina Agribition—grand champion, 5 of top 6 were Limousin. Toronto Royal Winter Fair—grand and reserve grand champion, 7 of top 10 were Limousin. Limousin heifers were grand and reserve champion carcasses with the top 4 out of 6 placings and the grand champion heifer was reserve grand champion carcass in the open competition against all steers including Limousin.

Limousin steer was supreme champion in the Queens Guineas at the Toronto Royal Winter Fair in November '81. Of the 257 entries in all breeds 92 were Limousin cross.

The first ever Indiana Field Day was held in early Sept. at the Limousin operation of Larry and Kathy Gerkin, Rockville, Indiana.

Gendarme, a French import and sire of many noted cattle in the United States and Canada, died at the age of 11.

Dec. 7—Grand champion female and top seller of the Black Limousin Futurity Show and Sale in Oklahoma was YKCC Miss Jackpot 817M. She sold for $21,000.

Feb.—In 1981, registrations of Limousin cattle were up 30% over 1980 in the United States.

The 12th annual National Limousin Sale at the National Western Stock Show in Denver, Colorado broke two breed records and one National Western Stock Show record.

Stretchs Genuine Risk, a fullblood daughter of MM Stretch, sold for $107,000 to Joli Farms, Bellvue, Washington and Mountain View Limousin, Enumclaw, Washington. The 1982 sale grossed $560,550 and averaged $9,921.

MK Sonic Jet, grand champion bull at the 1982 National Limousin sale brings $41,000 for a half interest. GKF Too Tall 13N, a fullblood Canadian Pacific daughter brings $40,000. TEXS Star Gigi, a fullblood Espoir de Carnaval daughter brings $31,500. "Movin' On," a fullblood MM Stretch son, brought $29,600.

Mar.—Jan. 23—One of the nation's premier operations, Hutchinson Limousin Farm, Thomson, Georgia, disperses. PURE Hermine 226J, sells to Hughes Cattle Company, Hugo, Oklahoma for $34,000. At the conclusion of the Hutchinson sale, semen was offered on many of the breed's leading sires. 91 units of Espoir de Carnaval semen sold for $600 per unit.

Long time Limousin breeder, William Schermer, Latimer, Iowa has been inducted into the Iowa Cattlemen's Hall of Fame.

Feb. 7—An era of Limousin history came to an end as the well known Nordic Farms herd, Campbellville, Ontario, dispersed. Nordic Katina, a full sister to Nordic Debonnaire, sold to E.J. Bishop, Winters, Texas for $26,000.

In recent months two major universities, Colorado State University and Oklahoma State University have announced plans for the development of Limousin herds within their animal science departments.

Alabama Limousin Breeders Association is formed.

Mar. 11—Symens Brothers, Amherst, South Dakota sells a flush, guaranteeing 4 embryos, in a double-black, polled daughter of SY Niterider for $23,000 in their annual production sale.

Feb. 22—Three females top $20,000 at the Annual Houston Show and Sale.

May—Apr. 4—An all black offering, the "Black Vintage Sale" saw 102 lots gross $265,370 with a sale top of $36,000 for a purebred November Innovator transplant calf and her dam.

June—Apr. 18—H.A. and Hazel McCoy of Seneca, Missouri disperse their Limousin herd and set breed record for a dispersal with a $9,640 average. Fleuralie, the dam of Budrick's Innovator, sold for $54,000. Miss McCoy Hillary 14L, the top selling heifer, brought $35,000.

A three-quarter interest in YKCC Michelin 904N brings $22,800 at annual production sale of the Yackley-CMC operation in Onida, South Dakota. One-tenth breeding interest in CMC Snowball demands $11,000.

Setting new price record at the 5th Annual Canadian Limousin Association Bull Test Sale at $27,000 was Bonaparte Norseman 114N. Consignment by Bonaparte Ranch, Cache Creek, British Columbia, the son of Pub, a French import, was purchased by Glenkair Farms, Grandview, Manitoba.

Worrell Land and Cattle Company, Charlottesville, Virginia has donated 17 Limousin breeding females and 13 calves to the Virginia Polytechnic Institute Animal Science Foundation in Blacksburg, Virginia.

Apr. 17—145 buyers from 15 states attended Jimmy and Betty Tinsley's 432 lot Limousin dispersal sale.

May 17—Two identical twin heifers, produced by non-surgically recovered embryos from a donor cow, splitting and transferring them to a recipient mother, are born at Colorado State University in Fort Collins, Colorado. These are the first in the world to be produced through non-surgical embryo recovery and transfer.

July/Aug.—Gene McKown, Norman, Oklahoma, has announced the sale of $92,500 worth of fullblood cows and a 10% breeding interest in the bull "Easy Jet" to Culpepper Limousin Company of Afton, Oklahoma.

May 15—North Texas Limousin Classic at the Bingham Ranch, Justin, Texas sees Nordic Jeanne XIM 34J sell for $50,000 to Joli Farms, Bellvue, Washington.

Grand champion steer of the 1982 San Angelo Stock Show and Rodeo was a fifteen-sixteenths Limousin steer.

Sept.—July 9 to 14—The Canadian Limousin World Congress was held in Calgary, Alberta. 253 head from Canada and United States were judged in a two day extravaganza by Berwyn Wise of Irricana, Alberta. High selling female of the Congress was Nordic Natina. Consigned by Cherry Lane Limousin of Parkhill, Ontario she sold to Joli Farms of Bellvue, Washington for $40,000. Top selling bull was the French import bull, Roustan, the show's reserve senior champion bull. Selling at $26,500 for a half interest to the Roustan Syndicate of Waterloo, Ontario, the son of Major was consigned by Golden Limousin Acres of Calgary, Alberta. 59½ lots averaged $8,763.

July 29 to 31—"Direction '82" Symposium was held in Wichita, Kansas to coincide with the National Junior Heifer Show.

Nov.—Grand champion 4-H market steer of 605 entries at the 55th annual Ak-Sar-Ben Livestock Exposition in Omaha, Nebraska was a Limousin cross steer.

Dec.—Mid-America Classic held in Marquez, Texas saw a fullblood Bold Type daughter with a 747 heifer calf at side sell for $25,400. A half interest in a Diplomat daughter brought $24,000.

Andy Grimes, Boaz, Alabama joins the *International Limousin Journal* staff as its eastern representative, replacing Carroll T. Cannon.

Seven Limousin embryos from Worrell Land and Cattle Company, Charlottesville, Virginia have been exported to South Korea's National Livestock Breeding Association.

Dr. Ray Hansen of Fort Collins, Colorado was killed in a small plane crash in Sioux Falls, South Dakota on November 30. Injured in the same crash was *International Limousin Journal* publisher Robert S. Crook.

1983

Jan.—Dec. 1—The Yackley-CMC "Genetic Designer" Sale in Sioux Falls, South Dakota set a new record average for a Limousin sale as 72 lots averaged $11,127. The top 20 females selling averaged over $24,000. High selling female was CMC Miss Gibraltar 133L selling for $50,500 to Ted Gibson, Winterville, Georgia. CMC Gendarme Gal 100L sold for $49,500. CMC Miss Hercules 135L sold for $40,000.

Dec. 20—The first shipment of cattle from Europe directly to United States soil have arrived at the Harry S. Truman Import Center in Florida. The 45 fullblood females and 5 fullblood bulls are owned by Horizon Limousin Ranch, Madisonville, Texas.

E. John Bruner and his wife Eileen, Winfred, South Dakota, were recognized as Seedstock Producers of the Year at the annual banquet of the South Dakota Beef Cattle Improvement Association in Chamberlain, South Dakota.

Prince Pompadour, one of the most popular Limousin bulls in history and the first fullblood Limousin animal imported by North American cattlemen in November of 1968, died in November in Riverside, California at the ranch of Limousin breeder Chet Edmunds. He was 15.

Feb.—Jan. 13—54 lots in the 13th annual National Limousin Show and Sale in Denver average $8,228, with six embryo transplants averaging $11,233. Top selling female of the sale was Bauchman Guy, the show's reserve grand champion female selling for $47,000 to Horizon Limousin Ranch, Madisonville, Texas. Also selling to Horizon was the grand champion bull, Bishop's Quarterback, selling for $40,000. Symens Brothers, Amherst, South Dakota had both the second high selling bull and female. The reserve grand champion bull, SYBB Dakota Innovator sold to Kenwynn Farms, Carseland, Alberta for $24,000. Dakota Pass 17N, the show's reserve senior champion female sold to Prine Limousin, Lithia, Florida for $22,000. Horizon Limousin Ranch purchased the high selling embryo sired by Harvest Olympus out of TEXS Star Fadette III for $25,000.

Yackley-CMC Limousin Cattle Company, Onida, South Dakota had the champion carload at the National Western Stock Show for the second year in a row.

Symens Brothers and Bain Limousin, Amherst, South Dakota had the champion pen of three at the National Western for the second year in a row.

Horizon Limousin Ranch, Madisonville, Texas and owner, Omar Bunkheila, bid $30,000 to help the North American Limousin Junior Association fund when a Limousin painting was auctioned off during the banquet at the National Western Stock Show.

According to NALF records more YKCC Jackpot progeny were recorded in 1982 than any other sire.

Jan. 14—10th annual Summit Sale held in Bennett, Colorado featured the dispersal of the Lazy Brook fullblood Limousin herd. Hillside Farms, Chauvin, Alberta, paid $20,000 for DH Precious LBLN 223N.

Apr.—Top selling bull at the Texas Limousin Association Show and Sale was the grand champion purebred bull, Bishop's Nate selling for $34,000. Bishops Nonie was the top selling female at $30,000.

Top selling bull at the Symens Brothers annual Production Sale in Amherst, South Dakota was SYBB Dakota Extra, a member of the Symens' champion pen of bulls from the National Western Stock show, selling to Bonaparte Ranch, Cache Creek, British Columbia for $42,000.

May—Top selling bull at the Schermer & Sons Annual Bull Sale was a purebred son of WS Visa 87J. ACH Farms, Flesherton, Ontario paid $25,000 for WS Cy 361M.

Effective August 1, 1983, the *International Limousin Journal* office will be permanently moved to Albany, Texas. Owner Bob Crook of Fort Collins, Colorado sold the ILJ to Buddy Sievers of Albany, Texas and Mike Merrell of Denver, Colorado in mid-April.

June—A fullblood son of Punch out of Columbia Houpette 4H, owned by Bob Dwyer and John Spivey of Georgia, broke the yearling record for a fullblood Limousin bull. He weighed 1,346 lb. on May 13th.

One of the Limousin breed's leading performance sires, Mr. LCCO, died in March 1983. He topped the first NALF Sire Summary in 1978 for weaning and yearling performance of his progeny. Mr. LCCO set an individual breed record himself for weaning, 869 lb., and yearling 1,444 lb., performance.

Apr. 19—YKCC Atlantic, a seven-eighths son of 747 sold for $43,000 for a three-quarter interest at the Yackley-CMC Cattle Company annual production sale in Onida, South Dakota. Little Mondeaux Limousin, Ogema, Wisconsin paid $24,000 for one-half interest in YKCC Chairman of the Board and $23,000 for YKCC Newsgirl.

McKinney-Richarz of Yoakum, Texas paid $40,000 for Ragora, an import daughter of Mimosa, at the Five Star Limousin Alliance dispersal in Gravette, Arkansas. McKown and Spitz, Norman, Oklahoma paid $31,100 for Regina, a daughter of Gardon.

July/Aug.—June 14-18—162 heifers are shown at the National Junior Heifer Show held in San Antonio, Texas.

Spitz and Sons Limousin, Oklahoma City, Oklahoma paid $22,000 for all of the embryos from one flush from a cow of his choice from the TEXS Limousin herd in New Braunfels, Texas.

One-half interest in Lonestar, a fullblood son of 747, sold to the Lonestar Syndicate for $50,000.

One-tenth interest in Cold Cash sold to Lowder Limousin, Canton, Texas and X Bar H, Houston, Texas for $11,000.

Consigned to the Atlantic Invitational Sale by McKown Limousin Ranch, Norman, Oklahoma, MH Overwhelming sold to Mike Hughes, Hugo, Oklahoma for $42,000.

New record for an embryo selling at auction was made when Grapner Farms, Barnesville, Georgia paid $40,500 for a Network x Deb's Lady embryo consigned by Bingham Limousin Ranch.

A new breed record was set at the Bishop Boys Honor Roll Female Sale, June 4, when Horizon Limousin Ranch, Madisonville, Texas paid $100,000 for half interest in Deux Amis Honey.

Oct.—As of the end of July, the NALF office has recorded 47,000 cattle, an increase of about 15 percent over last year and the largest number of cattle ever recorded by the Foundation in one year.

Volume 1, Number 1 of the *Limousin World* published by Dale Runnion is printed.

Goldenview Farms "Sale of the Century" averaged $9,028 on 74 lots. 15 cow/calf pairs averaged $14,437. Goldenview Lady GVC 233K sold to Dr. Vernon Ryan, San Angelo, Texas for $46,000. Goldenview Neda GVC 606N sold to Mountain View Limousin, Enumclaw, Washington and Elkhorn Ranch for $42,000.

One-half interest in BHLF Miss Hermine 1M sold for $25,000 to Prine Limousin, Lithia, Florida and CMC Corporation, McDonough, Georgia at the Pruitt Limousin 1st Production Sale in Homer, Georgia.

Nov.—Sept. 26—For the first time in the 56-year-history of the Ak-Sar-Ben Livestock Exposition a brother-sister duo showed the grand and reserve grand champion steers in the same year. James and Julie Symens, Amherst, South Dakota exhibited their Limousin crossbreds to championship honors over the 652 entries.

Mike Armitage, Claremore, Oklahoma has been named secretary-fieldman for the Oklahoma Limousin Breeders Association as of July 1983.

Burwell and Elaine Bates, who registered the first halfblood Limousin in Oklahoma dispersed their mature cow herd Sept. 10. The Real McCoy Ranch of Seneca, Missouri purchased the top selling female, Circle Honour for $51,000.

The first Brahmousin Field Day was held October 1 in Mountain Home, Texas at the 3 Mill Ranch.

John and Jackie Spivey, owners of CMC Corporation dispersed their percentage and purebred herds on Sept. 26.

Dec.—Oct. 10—A new breed record average of $16,045 was set on 42½ lots at the first Virginia Worrell Land and Cattle Company Sale. Mountain View Limousin, Enumclaw, Washington paid $92,000 for WLCC Forever, a daughter of 747 and Nordic Lafa XIM 74L. Miss Merva 18E sold to Nicholas Farms, Conyers, Georgia for $59,000.

Oct. 12—Hill Bright Farms disperses fullblood herd and Sunbelt Limousin disperses embryo donors. 49 females average $9,004.

Oct. 23—First offering of purebred Brahmousin sell in Cherokee Classic in Perry, Oklahoma.

1984

Jan.—Val Eberspacher of Beaver Crossing, Nebraska recently joined the *Limousin World* staff as a field representative for the Midwest.

Spitz, Ellis-Vantrease and Hansen Production Sale averages $6,851 on 129½ lots. One-half interest in Kiss Me Quick sold for $88,500 to Sunbelt Limousin, Suwanee, Georgia. CMC Miss Crystal sold to Horizon Limousin, Madisonville, Texas and Mountain View Limousin, Enumclaw, Washington for $104,000.

Feb.—Greg Martin, Executive Vice-President of NALF, has been appointed chairman of the Live Animal Evaluation Committee of the Beef Improvement Federation.

National Sale in Denver, Colorado set a new record for a National Sale average at $11,250 as well as a new record for a purebred female at auction. MK Black Hanni, a black purebred Black Mercedes daughter consigned by McKown Limousin, Norman, Oklahoma sold to Worrell Land & Cattle and Glenmore Farms of Virginia for $30,000. A purebred Canadian Pacific daughter consigned by Ox & W Limousin, Huron, South Dakota sold to Glenmore Farms of Keswick, Virginia for $27,000. TEXS Star 511P sold to Horizon Limousin for $62,000. WINN Pennicle sold to Wulf Limousin, Morris, Minnesota for $20,500. Yackley-CMC, Onida, South Dakota took top honors in the carload show for the third consecutive year. Symens Brothers of Amherst, South Dakota took grand champion honors in the Pen of Three competition for the fourth year in a row.

December 17, 1983—Bishop-Scott Big Splash sale featuring all 1983 transplant calves averaged $4,841 on 73 lots with top heifer calf selling for $27,500 to Horizon Limousin. The top ten heifer calves average $13,640.

Mar.—Robert H. "Bob" Williams of Covington, Georgia died from injuries received in a farm accident on January 14 at the age of 57. Williams was an owner of Twin Hills Farm, a partner in Sunbelt Limousin and a director of the North American Limousin Foundation.

The entire Bates Limousin cowherd of Konawa, Oklahoma, including 120 head of brood cows, 14 calves and the young herd sire, Trademark have been sold to Worrell Land & Cattle Co., Charlottesville, Virginia.

Omar Bunkheila of Horizon Limousin Ranch, Midway, Texas paid $12,000 to own a pair of boots donated to the National Junior Association from Mercedes Boot Co., Mercedes, Texas.

Pennsylvania Limousin Association is formed.

Apr.—Feb. 20—First National Brahmousin sale saw 48.95 lots of purebred and percentage Brahmousin average $4,244. 6 lots carrying the breeding of Cherokee Arauto 975 x Gloria averaged $24,442.

Feb. 12—CMCC Mr. Network, consigned by CMC Corporation, McDonough, Georgia and Quercus Farms, Gay, Georgia was grand champion bull and top seller at the Dixie National Show and Sale in Jackson, Mississippi. The Southeastern Syndicate paid $40,000 to own the 9/30/82 fullblood son of Network.

May—Mar. 12—Pub, a 2/4/79 French import fullblood son of Mimosa out of Jique, consigned by Limousin Sires of Canada, Dalemead, Alberta sold for $62,000. The half brother to Punch sold to Stewart Farming, Ltd. of Big Valley, Alberta.

Lee Pritchard has been named the Executive Secretary of the Texas Limousin Association.

Dave McNally, President of Canadian Cattle Consultants Ltd. (CCC), Calgary, Alberta announced that Cyril G. Hayes has sold his shares in CCC to Don Savage of Airdrie, Alberta.

200 Limousin steers were shown at the 1984 Houston Steer Show. The grand champion Limousin steer exhibited by Kody Newman, Stanton, Texas sold for $150,000 to Red Adair.

Circle J Ranch, Blenheim, Ontario paid $28,000 for the pick of the entire Bar-Dale Limousin female herd at Bar-Dale annual production sale, Stettler, Alberta. Circle J is a new firm owned by Chicago Cubs baseball pitcher, Ferguson Jenkins.

Miss Glenkair Helice, the dam of Manitoba Crocus and Canadian Pacific sold to the partnership of McKown Limousin, Norman, Oklahoma and Waddle Limousin, Platteville, Colorado for $34,000 at the annual Glenkair Farms production sale.

June/July—Bar-Dale Limousin and Bee-Jay Holdings, Erskine, Alberta gave $19,500 for a three-quarter interest in Dakota Chaparral, a 4/5/81 fullblood son of 747 at Symens Brothers annual production sale in Amherst, South Dakota.

The Dixieland Spectacular held in Atlanta, Georgia saw one-quarter interest in Nordic Hanni bring $52,000 which would put an evaluation of $208,000 on Hanni, a new breed record for a female. Double J Ranches, McCoy, Colorado purchased the embryo interest from Baker Limousin. Sweetwater Plantation, Thomson, Georgia gave $32,000 for one-half interest in 17 Hermine, a 1972 import daughter sired by 11 Elan. Straight Creek Farm, Fort Payne, Alabama gave $30,000 for a one-tenth breeding interest in Stretches Image.

Cressman Cattle Company and Mainline Limousin, Waterloo, Ontario purchased the high indexing bull at the Canadian Limousin Association Bull Test sale for $24,500. Glenkair Farms, Grandview, Manitoba consigned the fullblood son of Bonaparte Norseman.

Bates Limousin and Ellis-Vantrease Limousin dispersed the balance of their herds in Konawa, Oklahoma.

Dottie Robishaw, Houston, Texas gave $50,000 for Revanche, a four-year-old import daughter of Ocean out of Loanda consigned by Wynlane Ranch to the McKinney-Richarz Joint Production sale.

TEXS Royal Sale held at TEXS Limousin Ranch, New Braunfels, Texas averaged $11,030 on 36 fullblood females. TEXS Star Gee Whiz topped the sale at $60,000 selling to Woodhaven Plantation, Coolidge, Georgia.

The Michigan Junior Limousin Association held its inaugural meeting, February 18, 1984.

At the Yackley-CMC Limousin production sale, one-half interest in YKCC Apollo 243R sold to Dyer-Sims Ranch, Baird, Texas for $35,000. Angela Stewman gave $91,000 for Yellow Rose 034R a fullblood yearling by Colorado 153.

Horizon Limousin Ranch, Madisonville, Texas sold a one-tenth breeding interest in WRC Napoleon to Ryan Cattle Company, San Angelo, Texas for an undisclosed price.

40 head of fullblood Limousin from a dozen British breeders will be arriving at Horizon Limousin Ranch, Madisonville, Texas in June. The transaction was worth one quarter million pounds, ($335,000 U.S. dollars).

WLCC Second Thoughts consigned to the Annual Atlantic Invitational Limousin Classic sale by Worrell Land and Cattle Company, Charlottesville, Virginia sold to Bar H Ranch of Auburn, Georgia for $67,500.

Wayne Vanderwert has been appointed director of education and research for the North American Limousin Foundation.

Aug.—BCP Limousin Ranch, Big Hill, Texas paid $40,000 for one-half interest and one-half possession in TEXS Star Linda at Lazy J Ranch's Majestic Genetics Sale in Caldwell, Texas. Leo and Fran Marquette, Houston, Texas paid $25,000 for one-half interest in Spitz Gibraltar 151K and $12,000 for one-tenth interest in LJLR

Gold Card.

The 6th International Limousin Council meeting will be held in New Zealand in late October, early November.

Randy Bollum of Calgary, Alberta recently purchased the *Limousin Leader,* the Canadian Limousin breed publication. Bollum has been the editor of the publication for the past 4 years.

Juniors from 11 states have organized the Eastern Limousin Junior Association. This organization will help states east of the Mississippi get more juniors involved.

McKown Limousin Ranch, Norman, Oklahoma and Waddle Limousin Ranch, Platteville, Colorado have completed importation on 19 fullblood bulls and females from England and Scotland. Included in the purchases is Sirocco, a three-year-old son of Odeon.

Christopher Cliff and Richard Beale, Cheltenham, England, along with their wives, have formed Harvest Cattle Company, a buying service catering to North American cattlemen.

Beck Ranch of Summerland, British Columbia paid $37,000 for Goldenview Reva at the recent Highland Stock Farms Stampede Sale. Goldenview Farms, High River, Alberta gave $23,500 for Highland Raquel, (sired by Big Mac out of Highland Henrietta). 15 yearling fullblood females average $7,236.

Sept.—The *Limousin World* has been designated the official publication of the North American Limousin Foundation.

Spitz Limousin, Norman, Oklahoma has recently entered a sizeable embryo transplant program whereby most of the collected embryos will be frozen and stored. Initial plans are to freeze 600 to 800 embryos per year.

Limousin cattle will comprise a major portion of the more than 1,000 head of United States beef cattle to be exported to Libya.

North Dakota Junior Association is formed.

Oct.—Rick Dobson of Dobson Ranch, Marietta, Oklahoma recently announced the sale of Innovator to the X Bar H Ranch, Houston, Texas. The purchase price was not revealed.

Dec.—Worrell Land & Cattle WLCC Sale II averaged $18,894 on 48 lots to set a new breed record. One-half embryo interest and one-half possession in Nordic Lafa sold to Power Line Limousin, Bristol, Tennessee for $154,000. MH Gentle Thoughts sold with the same conditions for $100,000 to Power Line Limousin. WLCC Cover Girl sold to Horizon Limousin, Madisonville, Texas for $81,000. WLCC Peppermint sold to Woodhaven Plantation, Coolidge, Georgia for $58,000. Leo Payne, Big Timber, Montana paid $21,450 for 20 vials of 747 semen.

Horizon Limousin Ranch '84 Sale averaged $12,410 on 45½ lots. PRS Limousin, Houston, Texas paid $42,000 for an embryo flush out of Bewick Opal to set a breed record for an embryo flush. Opal is the record selling cow in England. Garland and Sydney Middendorf of Rocheport, Missouri paid $42,000 for an embryo flush out of Oakley Oona. Power Line Limousin, Bristol, Tennessee gave $24,000 for an embryo flush on their choice of females from Horizon's 1983 French importation.

Prime Properties Sale in Norman, Oklahoma saw 59⅔ lots average $14,037. One-half interest in Circle Jyntee sold for $215,000, setting a new breed record. Magness Limousin, Platteville, Colorado purchased Circle

Jyntee as well as paying $60,000 for MK Velvet Lady, $27,500 for SOLC Fancy Britches and $24,500 for MK Goldie.

Power Line Limousin of Bristol, Tennessee paid $37,000 for Cressman Jolie HMF 15J at Cressman's Prime Cut All Female Sale, making her the top selling female to sell at auction in Canada in 1984. The top 10 lots averaged $16,150.

Worrell Land & Cattle Co. of Charlottesville, Virginia paid $20,000 for a daughter of Eclair in Bonaparte Ranch's annual production sale.

Ox & W Limousin dispersed in a dynamic two-day dispersal which saw 839⅓ lots sell. One-third interest in Canadian Pacific sold for $11,000.

Yackley-CMC, Onida, South Dakta "Opportunity of the 80's" sale saw 415½ lots cross the auction block. Top selling lot was a purebred pair for $28,400.

1985

Jan.—George Stranahan, Woody Creek, Colorado purchased a one-third interest in Beyrand Joyeuse for $105,000 from Waddle Limousin, Platteville, Colorado in a private treaty sale, making Joy the second high selling female in the breed.

Feb.—Yackley-CMC, Onida, South Dakota has sold one-half interest and half possession of YKCC Michelin 904N to Payne Land & Cattle Company of Big Timber, Montana for $75,000.

Donald C. Matthews of Calgary, Alberta has been inducted into the Canadian Agricultural Hall of Fame. One of the principals of Highland Stock Farms, Don is a past president of the Canadian Limousin Association.

The National Show and Sale in Denver, Colorado averaged $11,615 breaking the record set last year. 9 lots sold for $25,000 or more. Horizon Limousin, Madisonville, Texas paid $100,000 for a 20% breeding interest in Spitz Navajo (747 x CMCC Snowflake). TEXS Star Dulce sold to Payne Land and Cattle Company for $55,000. Payne also gave $27,500 for TEXS Star Florida, and $25,000 for WLCC Patience. Little Mondeaux of Ogema, Wisconsin paid $51,000 for Lickety Split and $30,000 for TEXS Star Glorious.

Mar.—For the 4th straight year Yackley-CMC, Onida, South Dakota had the grand champion carload at the National Western Stock Show. Symens Brothers had the reserve grand champion carload for the second year in a row.

A pen of 5 Limousin-cross heifers belonging to Harold Rinehart, Highmore, South Dakota won grand champion honors in both the pen and individual carcass categories at the National Western Fed Beef Contest, over 125 other entries.

Apr.—Ron & Tony Simek of Little Mondeaux Limousin, Ogema, Wisconsin recently purchased one-half interest in Miss Crystal from Mountain View Limousin, Eagle Point, Oregon for $225,000 setting a new record for a Limousin female.

Feb. 20—Horizon Limousin Ranch held the first Limousin breeder hosted ballroom sale in Houston. 44 lots averaged a new breed record of $20,527. 21½ French females averaged $12,707; 13 U.S. females averaged $18,846 and 9½ English females averaged $40,526. Ryan Cattle Company, San Angelo, Texas paid $86,000 for half interest in Bewick Opal, an English import. Payne Land & Cattle Company of Big Timber, Montana paid $77,000 for half interest in the French import, Tangeante. X Bar H Ranch, Houston, Texas gave $73,000 for Hampton Tania and Garland

& Sydney Middendorf, Columbia, Missouri paid $60,000 for Topaze.

Destiny (747 x Dynasty) sold to Payne Land & Cattle Company, Big Timber, Montana for $62,000 at the annual Texas Limousin Association Show and Sale. Payne also gave $40,000 for half interest in YKCC Sue Ellen. Scott's Oak Grove gave $32,000 for MASH Americana.

Feb. 18—Jack & Vonnie Wyatt, Miami, Oklahoma; Coyote Hills Ranch, Chattanooga, Oklahoma; and Bingham Ranch, Justin, Texas paid $10,000 for a 5% semen interest in Tuxsing Super Star, a ¼ L x ¾ B bull, at the 2nd annual National Brahmousin Sale in Houston, Texas. This price put an evaluation of $200,000 on Super Star. A new record for a Brahmousin female was set when X Bar H Ranch, Houston, Texas paid $34,000 for one-half interest in DWIG 151K. Wynlane Ranch, Junction, Texas paid $33,500 for one-half interest in VRCC 706R. The 13 halfbloods averaged $7,065.

Andy Rest, western field representative for the *Limousin World* will wed Debra Wise, office manager for the Canadian Limousin Association August 10.

Magness Limousin, Platteville, Colorado paid $37,000 for Lana, an Enterprise daughter, at Greenwood Limousin's production sale in Lloydminster, Saskatchewan.

Saskatchewan Lily, a Helix daughter, sold to Magness Limousin, Platteville, Colorado for $40,000 in Guth Brothers' production sale in Lloydminster, Saskatchewan.

Grand champion steer at the San Angelo Stock Show was a Limousin cross steer shown by Suzanne Stewman, Maryneal, Texas selling for $46,000. Reserve grand champion steer was also a Limousin cross exhibited by Kody Newman, Stanton, Texas.

June Runnion has been appointed editor of the *Limousin World.*

May—Jim McKown & Sons, Springer, Oklahoma dispersed their well known Limousin herd March 13. Jim and his brother Gene gave the well known herd sire, Seven Forty Seven, his first breeding home.

In a private treaty transaction, Ron and Tony Simek of Little Mondeaux Limousin, Ogema Wisconsin purchased one-quarter interest in Pertinente for $250,000 from Jim and Ruth McBride of Goldenview Farms, High River, Alberta, putting a 1 million dollar evaluation on the Goldenview herd matron.

MK Lana (747 x Circle Jyntee) sold to Bingham Limousin, Justin, Texas for $97,000 at the Dixieland Spectacular in Atlanta, Georgia. Little Mondeaux Limousin, Ogema, Wisconsin and Horizon Limousin, Madisonville, Texas teamed up to purchase MK Split Ear for $44,000. Hagood Limousin, Auburn, Georgia paid $43,000 for Spitz Gibraltar 169K (Split Ear).

The Stewman family averaged $1,856 on 84 Limousin and Limousin cross club calves and $2,103 on 15 purebred show heifers at their 3rd annual club calf sale. Top selling steer brought $19,000.

June/July—At the Yackley-CMC Production Sale, Onida, South Dakota, half interest in 4 fullblood Crystal granddaughters by Punch averaged $17,125.

At the 8th annual Canadian Limousin Association Bull Test Sale in Calgary, Alberta a Bonaparte Norseman son was the top indexing and high selling bull for the second year in a row.

At TEXS Limousin's "Herd Builder" sale, a flush from TEXS Star Marquesa sold to Horizon Limousin, Houston, Texas for $23,000. Power Line Limousin gave $20,000 for a flush from TEXS Star Georgia Belle.

At the First Edition Sale hosted by Bingham Limousin, Justin, Texas, one-half embryo interest in WRC Deb's Lady sold to Bill Hagood, Auburn, Georgia for $70,000. Deb's Lady's production record showed her live progeny averaging over $14,000 and her embryos over $7,000.

Al Larson, owner of Mountain View Limousin suffered a fatal heart attack May 5 in Medford, Oregon.

Wes Ishmael of Longmont, Colorado has joined the staff of the *Limousin World* as the magazine's new editor. Ishmael has been a *World* intern for the past two years.

The West Virginia Limousin Association was formed April 28.

Dr. Wayne Vanderwert, Director of Education and Research for the North American Limousin Foundation was elected to a 3 year directorship of the Beef Improvement Federation.

Leonard Wulf, Morris, Minnesota was named the Minnesota Purebred Cattleman of the Year by the state's Beef Cattle Improvement Association.

Nordic Debonnaire (Email x Estouffe) made her North American presence felt as she and two daughters accounted for $328,000 in sales...half interest in Debonnaire for $58,000 to Payne Land & Cattle Co., Big Timber, Montana; full interest in Dynasty, a Network daughter for $200,000 to Payne Land & Cattle Co. and half interest in WRC Deb's Lady, a Gendarme daughter to Bill Hagood, Auburn, Georgia for $70,000.

Aug.—John Davison Moore, long time Limousin breeder from Newville, Pennsylvania died recently in a farm accident. John was a founder member and the second president of the North American Limousin Foundation.

Sept.—Juniors from 19 states converged in Springfield, Missouri for the largest ever national junior heifer show. 8 division champions, including the grand and reserve were by 7 different sires.

For the third consecutive year, Richard McKown, Norman, Oklahoma exhibited the grand champion female at the National Junior Heifer Show. No one except McKown has shown more than one national junior heifer show grand champion.

The first National Junior Brahmousin Heifer Show was held in Springfield, Missouri. Kristi and Keith Kent, Sweetwater, Texas exhibited grand and reserve grand females.

Dale and June Runnion have been made honorary members of the North American Limousin Junior Association in recognition for their years of dedication to the Limousin breed and support of the juniors.

170 head compete in the first All-American Limousin Futurity, July 19 in Springfield, Missouri. Spitz Navajo and Spitz Special Effort were named champions.

Choice of two yearling daughters of Pertinente, the million dollar cow, was the feature lot at Goldenview's Stampede Spectacular Sale in High River, Alberta. Waddle Limousin of Platteville, Colorado paid $40,000 for a 12/11/83 daughter of Harvest Olympus.

Yackley-CMC, Onida, South Dakota have joined forces with Little Mondeaux Limousin of Ogema, Wisconsin to form Yackley-CMC-Little Mondeaux Limousin based in Onida, South Dakota. CMC Limousin and Little Mondeaux will each continue their own individual operations. Bob Yackley will be the managing partner.

198

Jim Davidson of Shepherd, Montana and Bob Yackley of Yackley-CMC-Little Mondeaux, Onida, South Dakota have announced the sale of a majority of Davidson's Diamond V herd to the South Dakota program.

Pub, an import son of Mimosa is the leading Limousin sire in the Canadian Beef Sire Summary.

Oct.—Mountain View Limousin dispersed August 20 in Medford, Oregon. 6 donor females and recipient packages averaged $31,475.

Master Brun 279K and Explorer BUIT 1L have been chosen by American Breeders Service of DeForest, Wisconsin for use in a new European breeding program. Semen on both trait leaders in the 1985 NALF sire summary has been shipped to Holland.

Rio Vista International, Inc. has announced that they have developed a technique to successfully determine the sex of bovine embryos.

164 breeders from 19 states are currently registered with the North American Limousin Foundation as breeders of Brahmousin.

Nov.—Sept. 14—The Brahmousin Blowout held in Dallas, Texas sold 16.7 lots for an average of $24,839. Top selling lot was a half interest in a ½L x ½B Revolution daughter selling for $52,000.

Dec.—The III Virginia WLCC Sale averages $16,383 on 42 lots. One-half interest in WLCC Silhouette sold to Thomas Tallarico, Springfield, Massachusetts. Straight Creek Farm, Fort Payne, Alabama paid $62,000 for WRC Moon Maiden. Quercus Farms, Gay, Georgia gave $40,000 for one-half interest, one-half possession in WLCC M-X.

One-half interest in MK Stephanie sold to Magness Limousin, Platteville, Colorado for $100,000 in McKown's Prime Properties sale.

Vernon and Diana Ryan, San Angelo, Texas paid $40,600 for Druk Larissa at the British Cattle Society's Annual U.K. Breeders Sale held in Branbury, England and $37,800 for a two-year-old Favori daughter.

The 500,000th Limousin animal to be registered with the North American Limousin Foundation was a percentage female sired by Mr. Pompadour Hills 707P and owned by Erling & Linda Olsen, Dupree, South Dakota.

Dakota Pillar and Super Charge semen has been exported to the Republic of Panama for use on 100 selected Brahman females.

18.5 mature fullblood females at the Horizon '85 sale averaged $11,354. One-half interest in Oona, an English import sold to Little Mondeaux Limousin, Ogema, Wisconsin for $82,000.

Horizon Limousin, Madisonville, Texas gave $61,000 for one-half interest and one-half possession in the French import female, Ragora.

1986

Jan.—150 steers from 14 states have entered the North American Limousin Foundation's feedlot and carcass evaluation program, Lean On Limousin, in Creston, Iowa.

Feb.—The Southeastern Limousin Breeders Association recently announced the appointment of Randy G. Day as Association field representative.

A record 119 females and 40 bulls were shown at the National Western Stock Show Limousin show in Denver, Colorado. For the second year in a row, Spitz Limousin had grand champion female and bull. Spitz Navajo and Spitz Special Effort were also crowned winners of the Limousin Triple Crown.

At the National Limousin sale, MK Known Factor brought $50,500 for a 20 percent semen interest selling to Quercus Farms, Gay, Georgia; Joli Farms, Bellevue, Washington; and James Dyer, Fort Davis, Texas. A 20 percent semen interest in Captain Hook sold for $50,000 to the Happy Hookers Syndicate. MASH Americana, sold to Little Mondeaux Limousin for $51,000. Russell McCrady, Waxahachie, Texas paid $30,000 for OAKF Extremity.

For the 10th consecutive year, a Limousin cross steer was selected as grand champion carcass steer at the Canadian Western Agribition, Regina, Saskatchewan. Out of 54 entries, Limousin took the 12 top placings.

One of the largest private treaty transactions of fullblood females in recent years has taken place as Baker Limousin Ranch, Lindale, Texas purchased 30 head from the Bonaparte Ranch, Cache Creek, British Columbia.

Mar.—For the 5th year in a row, Yackley-CMC-Little Mondeaux Limousin exhibited the grand champion carload at the National Western Stock Show.

For the second year in a row, Limousin reigned supreme in the National Fed Beef Contest. This year, as last, the overall grand champion was a group of Limousin females.

Karl C. Dean, principal of Dean's Limousins, Inc., Salt Lake City, Utah, was named the state's seed stock producer of the year.

Apr.—Pam Beam of Liberty View Limousin, Wilmington, Ohio was named Young Cattleman of the Year by the Ohio Cattleman's Association.

The Horizon Ranch Shamrock Hilton Sale in Houston, Texas averaged $12,598 on 32.3 lots. One-half interest in Tambourine, an English import, sold to Perry Smith, Houston, Texas for $39,500. One-half interest in Fantasia sold to KK Cattle Company, New Braunfels, Texas for $35,000.

A flush from Goldies Victoria, the record high selling female in England, sold to Triple Creek Limousin, Wortham, Texas for $21,500 at Pinewood Farms first production sale in Houston, Texas.

Little Mondeaux Limousin, Ogema, Wisconsin and Jerry McManus, Humble, Texas gave $60,000 for MASH Surprise, the grand champion female at the Houston Limousin show.

Greenwood Shooting Star (MM Stretch x Lana) sold to High Low Limousin, Elk Point, Alberta for $20,000 at Greenwood Limousin's annual production sale.

The grand champion steer at the San Antonio Livestock Show was a Limousin cross steer that later sold for $40,000.

Dorothy Faidley, Colfax, Iowa has been named chairman of the Iowa Beef Industry Council. Faidley is a past president of the Limouselles.

May—The Hagood Limousin Dispersal in Auburn, Georgia averaged $11,281. One-half interest in WRC Deb's Lady sold to Payne Land & Cattle, Glenwood Springs, Colorado for $49,000. WLCC Second Thoughts sold to Rosanky Cattle Company for $31,500. WLCC Tomorrow sold for $30,000 for M&L Limousin, Era, Texas.

Worrell Land & Cattle Company, Charlottesville, Virginia paid $20,000 for a flush out of Gee Whiz at the Woodhaven Plantation Heritage Connection '86 Sale.

Limousin Breeders of Canada, Sherwood Park, Alberta and Molalla, Oregon paid $24,000 for Sirloin Princess at Stewart's Unsurpassed Sirloin 10th Production Sale in Big Valley, Alberta.

The grand champion Limousin steer at the Houston

show went on to be named the grand champion carcass animal over all breeds in the show.

June/July—One-half interest in OAKF Rainbow sold to Worrell Land and Cattle Company, Charlottesville, Virginia and Twin Hills Farm, Covington, Georgia for $50,000 at Quercus Farms "Better by Genetic Design" sale in Gay, Georgia. Little Mondeaux Limousin, Ogema, Wisconsin paid $40,500 for one-half interest in OAKF Iridescence. Dorminy Brothers Land and Cattle Company, Fitzgerald, Georgia paid $30,000 for a flush with 5 pregnancies out of WRC Missy.

Glenkair Farms, Grandview, Manitoba paid $20,000 for one-half interest and full possession in YKCL Zeus at the Yackley-CMC-Little Mondeaux Limousin production sale, Onida, South Dakota.

For the third year in a row the top indexing bull in the Canadian Limousin Association Bull Test was a Glenkair Farms bred son of Bonaparte Norseman, who was top indexing and top selling bull in the same test in 1982. Norseman sons have also been the top selling bulls for the last three years.

McKown Limousin, Norman, Oklahoma and Magness Limousin, Platteville, Colorado have formed a new partnership, Magness-McKown Cattle Company. The transaction includes 67 mature fullblood females, 17 syndicated bulls and 104 transplants.

Prine Limousin, Lithia, Florida have exported 3 fullblood Limousin bulls to Guatemala.

Coyote Hills Ranch, Chattanooga, Oklahoma dispersed 228 lots in a 2 day sale. 11¼ fullblood donor females averaged $13,307. 60 percent interest in Deb's Special Playmate sold to Little Mondeaux Limousin, Ogema, Wisconsin for $30,000. Little Mondeaux Limousin and Horizon Limousin, Madisonville, Texas paid $23,500 for 40 percent interest in K&D Belle Misty. Gurley Limousin, Byhalia, Mississippi, gave $15,000 for one-third interest in Lazy J Special.

A new yearling weight record of 1,434 pounds (adjusted) was set in the Limousin breed by Mighty Mo (Uranium x TEXS Star Gisela). Desperado (El Presidente x Miss Gold Emperor 503L) beat the new record by 14 pounds with an adjusted yearling weight of 1,448 pounds.

PJBB Polled Talisman 514T became one of the top selling polled bulls in the breed when ACH Farms, Markdale, Ontario gave $12,500 for half interest and full possession.

Aug.—Rosanky Cattle Company paid $30,000 for a flush out of Destiny at the Payne Land & Cattle Company "Prophecy Sale". Little Mondeaux Limousin gave $25,000 for a semen interest in the English bull Talent.

Half interest in Deb's Lady sold to M&L Limousin, Era, Texas at Bingham's 7th annual production sale near Justin, Texas. Little Mondeaux Limousin gave $15,000 for 20 percent interest in K&D Belle Misty.

The first Western Regional Junior Heifer Show is held in Weed, California.

Limousin Breeders Association of the Carolinas is formed.

Sept.—The largest National Junior Heifer Show to date was held in Louisville, Kentucky with 194 head from 19 states.

Oct.—The American Brahmousin Council has purchased the *Brahmousin Journal* from founder and publisher Mike Armitage. Mike Lange will serve as editor and field representative.

Dale and June Runnion have announced the sale of their interest in the *Limousin World, Inc.* to Dan and Nancy Wedman, Yukon, Oklahoma.

Nov.—National Beef Check-Off starts. $1 per head must be paid to state beef council or beef commission each time cattle are sold. Funds will be used for advertising, promotion, research and new product development.

Due to difficult time in the agriculture industry Limousin registrations are down 9.7% and transfers are down 30% from last year.

Sept. 12-21—100th Anniversary of the Limousin Herd Book and Seventh International Limousin Council meetings are held in the Limousin home land, Limoges, France.

Sept. 7—First annual membership meeting and field day of the West Virginia Limousin Breeders Association is held.

Sept. 8—First ever open show in the Brahmousin breed was held at the West Texas Fair in Abilene, Texas. 47 head participate.

Dec.—Worrell Land & Cattle Company's "Nonesuch" Sale averages $13,095 on 46 lots. WLCC Everlasting (Seven Forty Seven x Nordic Lafa) goes to Globescope Inc. of Toronto, Ontario for $62,000.

New record price for a black, polled, purebred Limousin female is set when NB Limousin and Jorgenson Limousin of Chowchilla, California pay $51,000 for WLCC Black Chanel (Seven Forty Seven x MK Black Hanni) in the Nonesuch sale.

Dave & Audrey Hatcher's X Bar H Ranch dispersed after nearly a decade in the Limousin breed. Emerson Limousin of Midlothian, Texas paid $29,000 for the English import female Pinsac Mamouche.

Longfield Farms "Something Special" Sale in Listowel, Ontario sees Nordic Prudence (Mastercharge x Nordic Lafa) sell for $31,000 to LBC and Associates of Molalla, Oregon.

Partnership of Prine-CMC Limousin of McDonough, Georgia disperses.

Oct. 5—One of the west's oldest operations, Hunt's Elkhorn Ranch owned by Pat Hunt of Cle Elum, Washington, disperses.

Final results have been released by NALF on the first phase of the Lean on Limousin National Feedlot and Carcass Evaluation program that included 148 head of purebred and percentage steers. This group posted an average of $1.47 in carcass value per day of age. The industry average was estimated at $1.19 to $1.25.

Flying Dog Ranch of Woody Creek, Colorado paid $20,000 for one-half interest in MK Ulrica (MK On A High x Saskatchewan Lily) in the annual Prime Properties Sale in Norman, Oklahoma. One-half interest in PURE Hermine 226J sold to Sherwood Farms of Chattanooga, Tennessee for $18,000.

Debbie Norton, long-time employee of American Cattle Services at Chattanooga, Oklahoma has accepted a position as marketing director for the Keystone Program in Topeka, Kansas. Kathy Brooks and Mark Smith will take over her advertising and cataloging duties.

1987

Jan.—NALJA members Kris Maurer of Sutherland, Iowa and Lynn Boyle of Dexter, Iowa are breed representatives and charter members of the newly formed Iowa Junior Beef Breeds Council.

2 Hanchon av Hagadal (911 Eclair x 117 Fanchon), one of the early cornerstone herd sires, was struck and killed by lightning.

Baker Limousin, Lindale, Texas dispersed on November 22. Nordic Hanni sells to Spitz Limousin Farms of Norman, Oklahoma for $43,000.

The Pete Carmichael & Family operation of Lookout Limousin from Timber Lake, South Dakota dispersed their mature cow herd on November 25. AMB Limousin of Grants Pass, Oregon purchased 92 head from this sale. The 14 sons and daughters of WRC Favorite Merger and Deruset Hillary commanded a $3,055 average.

Grand Champion steer at the North American International Livestock Exposition held in Louisville, Kentucky was a seven-eighths El Presidente son exhibited by Steve Herzog of Sealy, Texas.

Amerant Acres of Mount Brydges, Ontario sold the first ever North American Limousin bull into England. English cattleman John Greene purchased Amerant A-1, a fullblood Captain In Command son.

Feb.—The 1987 Sire Summary uses a new method of calculation which increases the accuracy of the weaning and yearling traits. The multiple trait analysis ties the weaning and yearling EPDs together.

Dennis Bauman, well-known cattleman from Spencer, Iowa has joined AMB Limousin of Grants Pass, Oregon as its general manager.

Louisiana Junior Limousin Association is formed.

Mar.—108 females and 49 bulls compete in the 1987 National Western Limousin Show. Cressman Sky High (x Roustan) exhibited by Cressman Cattle Company, Waterloo, Ontario and Taurus Breeders Service, Inc. of Lone Grove, Oklahoma is named Grand Champion Bull. Grand Champion Female honors went to LML American Dream exhibited by Little Mondeaux Limousin of Genoa, Nevada.

1987 National Limousin Sale sparks a $12,090 average on 41.5 lots. Desperado 5103T, the breed's yearling weight record holder sold to Little Mondeaux Limousin of Genoa, Nevada for $90,000 for one-half interest. He was consigned by Stewman Ranch of Maryneal, Texas. Cressman Sky High sells to Taurus Breeders, Inc. of Lone Grove, Oklahoma for $86,000 for one-half interest. Top selling female was TEXS Star Hanna, consigned by JCL Cattle Company, Welch, Oklahoma and Canadian Valley Ranch of Seminole, Oklahoma. She sold to Little Mondeaux Limousin for $60,000. The Premier Syndicate of Rosanky, Texas gave $30,000 for three-quarter interest in TEXS Ranger Premier, consigned by JCL Cattle Company and TEXS Limousin of New Braunfels, Texas.

The Pen and Carload Show at the 1987 National Western saw Symens Brothers of Amherst, South Dakota exhibit the grand champion pen of three with coming 2-year-olds. Yackley-CMC-Little Mondeaux Limousin (YK-CMC-LML) of Onida, South Dakota exhibited the reserve pen of three with yearling full brothers. Grand champion carload honors went to Leonard Wulf and Sons of Morris, Minnesota with a group of coming 2-year-olds. The yearling champions exhibited by YK-CMC-LML went on to be the reserve grand champion carload.

Leonard Wulf of Morris, Minnesota is named 1987 North American Limousin Foundation president.

Rosalie Smith of Smithton, Missouri has been elected as president of the Limouselles.

Wendell Geeslin, manager for Magness-McKown Cattle Company of Norman, Oklahoma is named Herdsman of the Year at the National Western, while Floyd McGown's TEXS Limousin program of New Braunfels, Texas was named Premier Breeder.

Cinde Schuppe of Iliff, Colorado is named the 1987 National Limousin Queen. Dale and June Runnion received NALF's Achievement Award at the Annual Banquet, January 15. This award pays tribute to their dedication to the development of the Limousin breed.

Bain Limousin of Frederick, South Dakota exhibited the 1987 National Western grand champion commercial replacement bred heifers over all breeds. The Bain crew also exhibited the grand champion pen of open commercial replacement heifers over all breeds.

Double J Ranches of McCoy, Colorado had four head on display in front of the Limousin Headquarters, the Denver Fairmont Hotel, during Stock Show. The hotel also served Limousin beef on the menu of the hotel restaurant, McGuires.

Limousin steer captured the grand champion carcass award at the Canadian Western Agribition held in Regina, Saskatchewan for the 11th consecutive year. The top 17 carcasses in the show were Limousin steers.

Pioneer breeder M.E. "Marvin" Singleton, Jr. passed away January 25 in Dallas, Texas. Marvin was recognized as a founder of the Texas Limousin Association.

J.V. "Lefty" Elliott of Fresno, California passed away January 21 at the age of 66. Lefty was a past NALF president and a founder of the Limousin breed on the West Coast.

Joe Lewis, noted livestock judge and Polled Hereford breeder of Larned, Kansas was killed in a freak vehicle accident January 17.

Apr.—Little Mondeaux Limousin of Genoa, Nevada paid $63,000 for 150 units of Orchard Sudan semen at the Horizon Limousin Ranch "Royale" Sale in Houston, Texas. MASH U-Bet (x Spitz Navajo) demanded $50,000 for one-half interest from a syndicate of five breeders. JLM Limousin of Genoa, Nevada paid $77,000 for one-half interest in Golden Lassie. The first five lots across the block averaged $40,000. 37 lots averaged $10,532.

The "TLA Show Heifer of the Year" was awarded to SSAM Official, exhibited by Sam Scott of Rising Star, Texas. Desperado 5103T was named the "TLA Show Bull of the Year." He was exhibited by Little Mondeaux Limousin of Genoa, Nevada and Stewman Ranch of Maryneal, Texas. These awards are presented by the Texas Limousin Association.

NALF has released a new promotional video entitled "Lean On Limousin" which focuses on the evolution of the breed as a leading producer of lean beef and its phenomenal growth in numbers and popularity in North America over the past 20 years.

The grand champion female at the 2nd annual International Polled Congress in Des Moines, Iowa was Polled Brook. Exhibited by Richard Schaefer of Greenfield, Iowa, she sold to ACH Farms of Markdale, Ontario for $11,600.

Hillside Farms of Halltown, Missouri paid $11,000 for the show's reserve grand champion female consigned by Wulf Limousin Farms of Morris, Minnesota. Diamond Lou Limousin of Cedar Rapids, Iowa gave $10,500 for the top selling polled bull consigned by Leonard Limousin of Holstein, Iowa.

The Saskatchewan Syndicate gave $18,400 for one-half interest and full possession of Canadian Rectifier at the Glenkair Farms' "Tribute To The Source" Sale in Brandon, Manitoba.

Saskatchewan Lilly III brought $23,000 at Greenwood Limousin's 6th annual production sale in Lloydminster, Saskatchewan, selling to Ernie Livingston of Vermillion, Alberta.

The 17th Annual Meeting of the Canadian Limousin Association was held in Brandon, Manitoba February 5-7.

NALF has compiled a comprehensive listing of breeders from around the country with Limousin and Limousin-cross feeder calves for sale and is distributing this *Feeder Calf Directory* to feedlots, order buyers and others upon request.

J. Yorga Farms of Flintoft, Saskatchewan and ACH Farms of Markdale, Ontario paid $10,700 for Polled Savory (x Polled Jolt) at the Trails End Farms dispersal in Stuart, Iowa.

May—The partnership of Yackley-CMC-Little Mondeaux Limousin will be dissolved May 27 and 28 in Onida, South Dakota in the ClaimStake Sale.

Longmont, Colorado will be the site for the 1987 National Junior Heifer Show and the All-American Limousin Futurity. NALF will sponsor a symposium entitled "Back To The Future" in conjunction with these two shows.

"April Fool's Day Flabbergast," the Worrell Land and Cattle V Sale, saw 121 lots sell for $1,003,200 and average $8,291. Vernon and Diana Ryan of Ryan Cattle Company, San Angelo, Texas purchased 13 lots for $379,450. MK Dana was top selling female at $90,000 selling to Ryan Cattle Company. Little Mondeaux Limousin of Genoa, Nevada paid $65,000 for one-interest in Nordic Lafa. Spitz Limousin Farms of Norman, Oklahoma teamed up with Ryan Cattle Company to purchase WLCC Timeless for $61,000. Little Mondeaux Limousin also bought CMC Miss Snowflake for $55,000.

Goldenview Farms, High River, Alberta sold Goldenview Ulysses (x Thunderbolt) to Symens Brothers of Amherst, South Dakota and Gavin Wall of Victoria, Australia for $13,000 in their March 31 bull sale. This purchase represents the first Australian-held interest in a North American Limousin bull.

Bob Prine of Prine Limousin, Lithia, Florida dispersed his cow herd on March 21.

The Canadian Limousin Association National Bull Test Sale celebrated its 10th anniversary March 30 in Calgary, Alberta. The top indexing fullblood bull and top indexing bull overall went to Marvin Latimer and family of Hayard Farms, Innisfail, Alberta with a son of Graham's Universal.

At the annual Symens Brothers Production Sale, polled bulls top the sale.

Members of the Georgia Limousin Association hit a Limousin grand slam at the Georgia Junior Steer Show in Athens, Georgia. In the 331 head open show, Limousin exhibitors lead out all four division champions, one reserve division champion and both grand champions. Grand Champion Steer was exhibited by Jeffery Haire of Colquitt, Georgia. The El Presidente son was bred in the Rosanky Cattle Company program in Rosanky, Texas. Jennifer Matthews of Lincolnton, Georgia exhibited the reserve grand champion steer, a son of Seven Forty Seven bred by Piedmont Farms of Garfield, Georgia.

March saw Dale and Carole Barclay of Bar-Dale Limousin host their 9th annual "Heart of the Herd" production sale and Wilbur and Melva Stewart of Big Valley, Alberta host the 11th Stewart Farming "Unsurpassed Sirloin" production sale.

March 7—The first Kentucky Beef Expo Limousin Show and Sale is held at the Kentucky State Fair and Exposition Center in Louisville, Kentucky.

Half interest in Spitz Chisolm 937U sells to Worrell Land and Cattle Company of Charlottesville, Virginia for $10,100 at the Spitz Limousin production sale April 4.

All of the 2-year-old bulls in the Bruner Limousin Genetic Power Sale held in Madison, South Dakota March 25 sold with EPD values.

Phase II of the Lean On Limousin National Feedlot and Carcass Evaluation Program has been completed. 33 purebred and 16 percentage Limousin steers along with 10 Brahmousin steers were entered in the test. The average carcass value per day of age (CDVA) was $1.42 on the purebred and percentage Limousin steers. The beef industry average CDVA was estimated at $1.19-$1.25.

Tom and Rosalie Smith of Smithton, Missouri were presented with the highly coveted Outstanding Service Award from the Missouri Limousin Association.

O'Brien Limousin hosts their 13th annual bull sale March 15.

Shari Holloway of Chattanooga, Oklahoma exhibited the grand champion steer at the Wichita Falls Junior Beef Show in Wichita Falls, Texas with a fullblood son of Captain In Command. 161 steers were entered in this open show.

Dale Runnion has been selected to receive the Livestock Publications Council's (LPC) Headliner Award, the highest honor that can be bestowed by the LPC.

Nordic Debonnaire XIM 11D, the well known Email daughter out of Estouffe, died at Rio Vista Farms near San Antonio, Texas.

ACKNOWLEDGMENTS

The publishers and editors of *The History of Limousin in North America* wish to express our appreciation to Louis de Neuville, Limoges, France for the opportunity to participate in this project. His dedication to the completion of this account inspired us to continue at times when the task seemed insurmountable.

We are particularly indebted to Dick Goff, the author of Part I, and to Greg Martin, author of Part III. Bob Vantrease contributed documents and experiences along with his unfailing support for Part II. Several quotes and events in the book were supplied by Debi Robbe.

Laura Palmer, Harvey Tedford and the Canadian Limousin Association gave whole hearted cooperation and free access to files for compiling the Canadian Limousin History. Randy Bollum, the *Limousin Leader,* furnished us with most of the photos for this section. We are indebted to Debbie Wise Rest for her assistance with this portion of the book.

We especially thank Floyd McGown who so skillfully compiled our Foreword and advised us in direction and critiqued some of our work.

Production would have been impossible without the cooperation of the *Limousin World* owners, Dan and Nancy Wedman, who granted us the use of their files. We thank the *Limousin World* staff, Pamela Clifford, Lisa Cole, Kathaline Cook, Wendy Dickson, Wes Ishmael, Louise Kello and Debbie Rest. The quality attained is a tribute to their expertise.

ART CREDITS

Lisa E. Cole
Galen Struve
Jerry T. Sutton

Index